Corporate Social Responsibility in Management and Engineering

RIVER PUBLISHERS SERIES IN MANAGEMENT SCIENCES AND ENGINEERING

Series Editors

CAROLINA MACHADO
University of Minho
Portugal

JOÃO PAULO DAVIM
University of Aveiro
Portugal

Indexing: All books published in this series are submitted to the Web of Science Book Citation Index (BkCI), to CrossRef and to Google Scholar.

The "River Publishers Series in Management Sciences and Engineering" looks to publish high quality books on management sciences and engineering. Providing discussion and the exchange of information on principles, strategies, models, techniques, methodologies and applications of management sciences and engineering in the field of industry, commerce and services, it aims to communicate the latest developments and thinking on the management subject world-wide. It seeks to link management sciences and engineering disciplines to promote sustainable development, highlighting cultural and geographic diversity in studies of human resource management and engineering and uses that have a special impact on organizational communications, change processes and work practices, reflecting the diversity of societal and infrastructural conditions.

The main aim of this book series is to provide channel of communication to disseminate knowledge between academics/researchers and managers. This series can serve as a useful reference for academics, researchers, managers, engineers, andother professionals in related matters with management sciences and engineering.

Books published in the series include research monographs, edited volumes, handbooks and text books. The books provide professionals, researchers, educators, and advanced students in the field with an invaluable insight into the latest research and developments.

Topics covered in the series include, but are by no means restricted to the following:

- Human Resources Management
- Culture and Organisational Behaviour
- Higher Education for Sustainability
- SME Management
- Strategic Management
- Entrepreneurship and Business Strategy
- Interdisciplinary Management
- Management and Engineering Education
- Knowledge Management
- Operations Strategy and Planning
- Sustainable Management and Engineering
- Production and Industrial Engineering
- Materials and Manufacturing Processes
- Manufacturing Engineering
- Interdisciplinary Engineering

For a list of other books in this series, visit www.riverpublishers.com

Corporate Social Responsibility in Management and Engineering

Editors

Carolina Machado

University of Minho
Portugal

João Paulo Davim

University of Aveiro
Portugal

LONDON AND NEW YORK

Published 2018 by River Publishers
River Publishers
Alsbjergvej 10, 9260 Gistrup, Denmark
www.riverpublishers.com

Distributed exclusively by Routledge
4 Park Square, Milton Park, Abingdon, Oxon OX14 4RN
605 Third Avenue, New York, NY 10158

First published in paperback 2024

Corporate Social Responsibility in Management and Engineering / by
Carolina Machado, João Paulo Davim.

Routledge is an imprint of the Taylor & Francis Group, an informa business

Publisher's Note
The publisher has gone to great lengths to ensure the quality of this reprint but
points out that some imperfections in the original copies may be apparent.

While every effort is made to provide dependable information, the
publisher, authors, and editors cannot be held responsible for any errors
or omissions.

ISBN: 978-87-93609-61-7 (hbk)
ISBN: 978-87-7004-399-1 (pbk)
ISBN: 978-1-003-33773-7 (ebk)

DOI: 10.1201/9781003337737

Contents

2   Future-Focused Entrepreneurship Assessment (FFEA)   31
Niko Roorda

**3 Corporate Social Responsibility: The Case of East Timor
Multinationals**   **99**

Carla Freire, Manuel Brito and Iris Barbosa

**4 Gender Diversity and Equality in the Boardroom:
Impacts of Gender Quota Implementation in Portugal**   **147**

Mara Sousa and Maria João Santos

Preface

Integrated in a highly competitive environment nowadays organizations need to develop proactive strategies in order to find the best solutions to solve the serious problems that result from the human been evolution. Never than now organizations need to be responsible for its actions in the markets where they are present, leading us to new concepts such as Corporate Social Responsibility (CSR). Referring to organizations responsibility for their impact on society, CSR is greatly relevant for the competitiveness, sustainability and innovation in management and engineering arena of the organizations and the economy worldwide. Relevant to enterprises, the economy and the society as a whole, CSR brings important benefits for cost savings, customer relationships, risk management as well as human resource management, making organizations more sustainable, creative and innovative. It appears as a critical and recent management strategy under which organizations look to establish a positive impact on society at the same time that they are doing their businesses. Looking both to the management of its HR and processes improvement and to its impact on society, the organization stakeholders are greatly concerned with the activities that are developed and its impact in the environment and the society as a whole. The organizations management and engineering areas have here an important role as they need to act ethically, contributing to the organizational development at the same time they need to improve the workforce (and their families) as well as the society quality of life.

Taking into account these concerns, this book looks to cover the issues related to corporate social responsibility in management and engineering in a context where organizations are facing, day after day, high challenges in what concerns the items related to their social responsibility. It looks to contribute to the exchange of experiences and perspectives about the state of the research related to CSR, as well as the future direction of this field of research. It looks to provide a support to academics and researchers, as well as those that operating in the management field need to deal with policies and strategies related to CSR.

This book is designed to increase the knowledge and effectiveness of all those that are interested in develop a management system that looks to meet the needs of a transforming and responsible organization, in all kind of organizations and activity sectors.

Aiming to share knowledge, research results and experience, this book covers corporate social responsibility in management and engineering in six chapters. Chapter 1 discusses "The Boundaries of Corporate Social Responsibility: A Managerial Perspective". Chapter 2 covers "Future-Focused Entrepreneurship Assessment (FFEA)". Chapter 3 descrives "Corporate Social Responsibility: The Case of East Timor multinationals". Chapter 4 contains information on "Gender Diversity and Equality in the Boardroom: Impacts of Gender Quota Implementation in Portugal". Subsequently, Chapter 5 covers "Reconstructuring CSR in the Construction Industry". Finally, in Chapter 6 "Work-Family Conciliation Policies: Answering to Corporate Social Responsibility – A Case Study" is presented.

Whatever the type of professional we are, all of us need to know what is happening in the most diverse environments in order to understand and develop effective responses to meet all these new demands and challenges that we are facing at the present days.

This is the reason why the interest in this subject is evident for many types of organizations – namely, important Institutes and Universities all over the world –, as well as for a diverse pool of professionals, such as, HR managers, managers, engineers, entrepreneurs, strategists, practitioners, academics, researchers, among others.

The Editors acknowledge their gratitude to RIVER Publishers for this opportunity and for their professional support. Finally, we would like to thank to all chapter authors for their interest and availability to work on this project.

Carolina Machado
Braga, Portugal

João Paulo Davim
Aveiro, Portugal

List of Contributors

Adriana Faria *Department of Management, School of Economics and Management, University of Minho, Braga, Portugal*

Carla Freire *Department of Management, School of Economics and Management, University of Minho, Braga, Portugal*

Carolina Feliciana Machado *Department of Management, School of Economics and Management, University of Minho, Braga, Portugal*

David Starr-Glass *Center for International Programs (Prague Unit), State University of New York, Empire State College, USA*

Íris Barbosa *Department of Management, School of Economics and Management, University of Minho, Braga, Portugal*

Kwasi Dartey-Baah *Department of Organization and Human Resource Management, University of Ghana Business School, Accra, Ghana*

Kwesi Amponsah-Tawiah *Department of Organization and Human Resource Management, University of Ghana Business School, Accra, Ghana*

Manuel Brito *Department of Management, School of Economics and Management, University of Minho, Braga, Portugal*

Mara Sousa *School of Economics and Management, University of Lisbon, Lisboa, Portugal*

Maria Jõao Nicolau dos Santos *School of Economics and Management, University of Lisbon, Lisboa, Portugal*

Niko Roorda *Roorda Sustainability, Department of Learning and Innovation, Avans University, Tilburg, The Netherlands*

Yaw A. Debrah *School of Management, Swansea University, Swansea, UK*

List of Figures

List of Tables

1

The Boundaries of Corporate Social Responsibility: A Managerial Perspective

David Starr-Glass

Center for International Programs (Prague Unit), State University
of New York, Empire State College, USA

Abstract

Corporate social responsibility is a well-established but deeply contentious area of study. In its history, it has variously advocated that corporate behavior should produce minimal social harm or result in added-value for the multiple stakeholders that the corporation impacts. This chapter assumes a managerial perspective, that is, one located inside the corporation and looking out toward the economic and social worlds in which it operates. It considers the critical non-human persona of the corporation, the social and ethical actors with which it engages, and the resulting confusion and misunderstanding regarding the games that are being played. It sees the corporate and social worlds as disconnected and mutually invisible to game players and suggests that corporate social responsibility can have no intelligible meaning, let alone positive outcomes, until these separated worlds are brought together through dialog.

1.1 Introduction

In 1917, Henry Ford articulated what most people today would recognize as a laudable and characteristically pragmatic commitment to corporate social responsibility (CSR). At the time, Ford was being sued by his shareholders for reinvesting all of the corporate profits in the business and refusing to pay out any dividends. His intent was to optimize corporate growth, increase manufacturing capacity, and produce even more cars even more cheaply. When challenged by shareholders, Ford argued that the corporation should "do as much as possible for *everybody* concerned" and that the role of the corporation was "to make money and use it, give employment, and send out the car where the people can use it ... and *incidentally* to make money" [1, p. 100, emphasis added].

These were considered radical thoughts for the time and, in due course, his litigious shareholders prevailed in the courts and Ford was forced to limit his reinvestment strategy and pay out substantial dividends to his short-sighted shareholders. Clearly, Ford considered that the rules of the corporate game were more expansive than short-term profit maximization and wealth accumulation for corporate shareholders. This understanding of the role of the corporation contradicted the received economic and commercial wisdom of 1917, which might have been more accurately stated as: "the purpose of a company was *not to do as much good as possible*, but to make profit" [2, p. 171, emphasis added].

The question of whether corporations should do as much good as possible for everybody concerned has been asked with increasing urgency, and arguably with growing frustration, over the last hundred years – particularly by those outside the corporate world. Those inside that world often considered the question spurious because it seems obvious to them that the corporation has little or no social responsibility. Milton Friedman, advocating a neoliberal economic interpretation of the firm and relying on theories of agency and fiduciary duty, memorably reframed the question. He argued that "there is one and only one social responsibility of business ... to use its resources and engage in activities designed to increase its profits" [3, p. 133]. Interestingly, Friedman qualified this statement with a crucial but much less quoted condition – "so long as it [the firm] stays within the *rules of the game*, which is to say, engages in open and free competition without deception or fraud" [3, p. 133, emphasis added].

Multiple understandings, theories, and justifications have been advanced to clarify the assumed rules, participating players, and expected outcomes of "the game," but there have been fewer considerations of *which game* is in question or whether *multiple games* might be simultaneously in play. Clarity, however, has never been a significant element of the CSR discourse and its shifting and nebulous definitions have done little to help [4]. Many view CSR as a game in which the players, rules, and outcomes are best seen from a political perspective and where meaning emerges only through a dialog between the corporate world and civil society – a dialog that all too often degenerates into self-serving monologs [5]. Meaningful dialogs need to acknowledge the significant power differentials that exist and recognize "the specter of power asymmetries and the inevitability of conflict in stakeholder relations, particularly for powerless stakeholders" [6, p. 1]. Power differentials are certainly evident, but perhaps more obvious are the ideological differences between those involved – CSR clearly means different games to different people and little common ground can be reached until the competing narratives recognize, respect, and reconcile those differences [4, 7–9].

Sixty-five years ago, Bowen – recognized as the originator of the CSR movement – argued that there was an *obligation* for corporate leaders "to pursue those policies, to make those decisions or follow those lines of action which are desirable in terms of the objectives and values of society" [10, p. 6]. His argument gained little traction, particularly in the corporate world. Today, after the devastating financial tsunami of 2007 and the protracted Great Recession that followed, many are less concerned with Bowen's desirable lines of action and more focused on blatant corporate social *irresponsibility* [11–13]. Some, who previously had great confidence in the economic game, have been forced to reconsider their assumptions. In hearings before the US Congressional Committee on Oversight and Government Reform (October 23, 2008), even former Federal Reserve chairman Alan Greenspan – an eloquent and at times irrationally exuberant high priest of neoliberalism – admitted that he had "made a *mistake* in presuming that the self-interests of organizations, specifically banks and others, were such that they were best capable of protecting their own shareholders and their equity in the firms" (emphasis added).

If corporate *self-interest* cannot protect those who *are* the corporation (its shareholders), how can it be expected to protect anyone else who lies beyond the corporate boundaries? Some have advocated a greater sense of *corporate citizenship,* in which the business firm "can and – judged by the criterion of prudent self-interest – 'should' take on an active role in rule-finding discourses and rule-setting processes with the intent of realizing a win–win outcome of the economic game" [14, p. 375]. But, with corporate self-interest so narcissistically defined and pursued, it is unlikely that the firm's notion of win–win outcomes will align with the expectations of the non-corporate participants in that economic game.

This chapter explores the contested landscape of CSR. In doing so, and given the anticipated readership of this book, it assumes a managerial perspective – that is, a perspective located inside the corporation and looking out toward the economic and social worlds beyond. The chapter examines the challenges and opportunities that CSR provides by focusing on the different and contradictory meanings that the construct has assumed and the different and contradictory games involved. It might be anticipated that a managerial perspective would support the corporation's preferential concern for its most powerful stakeholders (putatively its shareholders) rather than its weaker ones (particularly employees and consumers), but this chapter does not promote a conventional business case approach. Instead, it argues that it will be increasingly necessary for corporations to rethink their neoliberal economic models and to reconsider their impact on the less powerful in society.

The chapter is constructed as follows. The next section sets the scene by considering the uniqueness of the corporation, which is legally endowed with a set of financial and economic advantages. Section 1.3 explores the possibilities of dualisms and dilemmas when the corporation, as a non-human actor, engages in the surrounding social and economic worlds. Section 1.4 continues this exploration by looking at the economic, social, and environmental engagement of the corporation and the possible trilemmas and tradeoffs that might arise in its encounter with these different spheres. Section 1.5 deals with the corporation's boundaries regarding economic, social, and environmental responsibilities and argues for a positive and pervasive corporate responsiveness, especially toward its least powerful and least influential stakeholders. Section 1.6 restates some of the major themes of the chapter and suggests ways in which CSR might become more of a pressing reality and a less of a rhetorical accommodation.

1.2 The Singularity of Corporations

Corporate social responsibility cannot be appreciated unless there is an understanding of the uniqueness of corporations, the reasons that they were created, and the behaviors that they were programmed to exhibit. Corporations – like present-day robots and cyborgs – are non-human beings created by humans. Corporations are a common feature of our world and we might believe that we have gained some familiarity with them. However, their creation and continuing existence should not be taken for granted because they pose as many questions – and raise as many concerns – as present-day driverless cars and artificial intelligence.

The modern corporation can trace its origins back to a remarkably innovative and creative construction of Roman law that took place almost two millennia ago. This audacious legal formulation allowed, for the first time, social collectives and groups of individuals to be regarded as independent non-human entities. These newly formed legal entities were afforded most of the legal rights and obligation of natural persons but – spectacularly – they were also given unique rights unavailable to the individuals who formed them. The underlying legal justifications for corporate creation were somewhat vague, applied in different ways in different contexts, and often appeared to be self-contradictory. However, this innovative legal formulation proved exceptionally useful for the non-human entities that were brought into existence and for the state, particularly in terms of extending its tax base and guaranteeing continuing tax revenues.

After the fall of Rome (476 C. E.), the corporate concept was enthusiastically take up by Byzantium and the rapidly expanding Christian church, and was subsequently elaborated and refined in the ecclesiastical courts of Medieval Europe [15]. Originally, corporate status was conferred on enterprises that had been created to provide public, communal, or social services but in time it was also seen as a highly advantageous form for business-minded entrepreneurs and merchant associations. Remnants of this evolving historical progression are reflected in our contemporary world – the oldest surviving examples are the ancient *universities* of Europe that date back to the Middle Ages (Latin: *universitas* = all turned into one); their relatives are our present-day business *corporations* (Latin: *corpus* = a single body). The forms, activities, and preoccupations of universities and business corporations seem remarkably different but both have the same legal origin. This commonality of origin was underscored by Pollock and Maitland – the preeminent jurists of 19th century England – who noted that commercial corporations were the linear descendants of those "corporations of one small class, the learned corporations that were founded in the 12th and 13th centuries, and others that in later days were fashioned after their likeness" [16, p. 459].

1.2.1 Unique Attributes of Corporations

The form and nature of the corporation are *legally* determined and the advantages gained are legal ones, even though they have profound social and economic consequences. In order to understand why the corporate form might be of particular economic importance – and why it might provide great instrumental value in the economic game – it is necessary to consider its fundamental legal characteristics. Specifically, corporate formation is designed to afford three guarantees of protection generally unavailable to other social and economic actors.

- *Shielding corporate owners:* Those who seek incorporation of their entity – individuals or groups (shareholders) – are provided with a legally recognized shield that protects them from the subsequent behavior and action of the newly formed entity. Although natural persons will participate in corporate affairs, when doing so they are generally separated, distanced, and protected from the legal consequences of the actions of the corporate entity. In particular, the claims of the firm's creditors are limited to the asset pool held and controlled by the corporation, not to the individual assets of its owners. Limited liability – a spectacular advantage of the corporation – is "a form of *owner shielding* that [operates] by protecting personal assets of firm owners from the claims of firm creditors" [17, p. 1336, emphasis in original]. Shielding the personal non-corporate assets of individual owners provides unsurpassed advantages for those who wish to create corporations because it reduces their risk exposure, protects their personal wealth, and makes the corporation's future capital acquisition easier. All of these significant advantages provided by the corporate form led to its popularity and dramatic growth, particularly in North America in the latter half of the 19th century [18].
- *Shielding corporate entities:* In many ways, entity shielding is the inverse of owner shielding. Owner shielding protects the personal assets of corporate owners from the claims of corporate creditors; whereas, *entity shielding* protects the *corporate asset pool* from the claims that creditors may have on the personal assets of those who form the corporation. Again, this provides asymmetrical benefits and advantages for corporate and personal creditors in the liquidation of the corporation. However, although the asymmetry produced by owner shielding can also be created through individual contractual arrangements, "it would be nearly impossible to develop effective *entity shielding* without special rules of law…[because] entity shielding limits the rights of personal creditors by subordinating their claims on firm assets to those of firm creditors, and strong entity shielding additionally limits their ability to liquidate firm assets" [17, p. 1338, emphasis added].
- *Continuing corporate life:* "A separate *indivisible legal personality* for the corporation ensures that it has a life of its own, and does not have to be broken up (and reconstituted) if any of its owners or employees die or leave" [19, p. 1188, emphasis in original]. Once formed, the corporation is granted what is equivalent to *legal immortality* that allows it to outlive the natural persons who formed it and who constitute its membership.

Legally granted immortality has profound and highly desirable consequences for the corporation as an entity and for its shareholders. The corporate entity, with the assurance of an indefinite and unrestricted future, can engage in continuous restructuring and future-orientated strategical planning. It can confidently embark on long-term enterprise building, engage in uninterrupted wealth accumulation, and concentrate on future-orientated economic growth that would be unreasonable, if not impossible, for natural persons.

All of these elements constitute very real and significant advantages for the corporate entity and for those natural persons who create and own it. These advantages are provided within legal frameworks – nominally connected to and endorsed by the state – that clearly recognize that there are substantial legal advantages to be gained from such forms of business organizations. What is less clear, however, are: (a) why these specific legal advantages should have been granted in the first place; (b) why the corporation, as a non-human entity, has been privileged over natural persons; and (c) what is the rationale for the resulting tradeoffs and competing claims that surround corporate creation?

The answers to these questions are matters of conjecture. Yet, it is obvious that the resulting non-human entities that have been created by the law will likely engage in social and economic transactions with natural persons and, when they do, they will be able to play a *legal game* that has significantly different rules from the ones that constrain those natural persons. The corporation has built-in advantages that not only differentiate it from human actors but also provide it with the ability to engage in economic activities – notably asymmetrical risk exposure, future-orientated wealth creation, and unlimited existence – that are not available to other economic players or stakeholders. Thus, the corporation will be able to play a different *economic game* that the one in which human actors engage.

1.2.2 Fictitious Entity or Contractual Nexus?

The paradox is that the fictitious corporate entity is very real. The corporate entity may be a fiction but as Lord Sumption, Justice of the U.K. Supreme Court, recently remarked: "the separate personality and property of a company is sometimes described as a fiction, and in a sense it is. But the fiction is the whole foundation of English company and insolvency law" [quoted in 20, p. 315]. The fiction is perpetuated and reified through the creation of the *corporate veil* that separates, shields, and distances corporate entities from natural persons. The corporate veil is a fiction, but it has a very material substance that is recognized, sustained, and rarely pierced – certainly in the legal systems of the English speaking world – and which leads to corporations operating as powerful and privileged players in the social and economic worlds [21–24].

An *entity* is something that is distinctive, possesses its own characteristics, and is identifiable from other similar things. The corporation has a legal identity and a legal existence; however, a number of legal scholars and economists do not believe that the corporation is a best described as a separate entity, no matter how immaterial. For them, the corporation is better regarded as a *nexus of contracts* [25] and the firm better seen not as "a thing, but rather but rather a nexus or web of explicit and implicit contracts establishing rights and obligations among the various inputs making up the firm" [26, p. 485]. Contract-centered perspectives understand that the corporation has no independent obligations or responsibilities, apart from those present in the nexus, because it lacks the status of an entity – it lacks an identity, a persona, a presence, and the ability to relate with the others. The corporation is thus no more than a "nexus for a mass of contracts which various individuals have voluntarily entered into for their mutual benefit. ... [and] is incapable of having social or moral obligations much in the same way that inanimate objects are incapable of having these obligations" [27, p. 1273].

Whether viewed as an independent legal entity or as a nexus of contracts, the corporation is a non-human presence in the social and economic spheres. Corporations exhibit a singularity that sets them apart from natural persons and they are able to engage in behaviors that would not be possible for those natural persons. Corporations are capable of playing different legal and economic games and it seems inevitable that they will play these games with rules and expected outcomes that differ from the games played by natural persons. In considering socially responsible behavior, it is important to appreciate that corporations are not human actors, social beings, or moral agents and that they are not – and cannot be – direct players in the social responsibility game. It is equally important to appreciate that corporations were specifically created to be different from natural persons and that their legally granted features give them the ability – the *instrumental capacity* – to behave in ways that are non-human. These features make them particularly well-designed to participate as privileged players in the game of *their* choosing – in the game in which they have an advantage and in which they can excel: the game of the neoliberal marketplace.

1.3 Dualisms and Dilemmas

Corporations have been purposefully created in order to protect their existences, perpetuate those existences, and provide a set of privileges and advantages that are significantly different from the rights and obligations of natural persons. However, although corporations are legal-privileged entities, they must eventually participate in the social and economic world that surrounds them. In that engagement: (a) will corporations exhibit *legal and economic behaviors* and act in ways that recognize their fiduciary duties to the few (shareholders), behaviors that will inevitably set them *apart* from the ambient social and economic worlds? or (b) will they exhibit *social behaviors* that recognize the possible moral duties and ethical obligations that they might have toward human actors and stakeholders (employees and customers), behaviors that will signal that they are *a part* of the broader social world?

These questions reflect the two different games into which corporations might enter. If these two possibilities exist, and if they are equally possible and amenable to choice, corporations would be confronted with a behavioral dilemma. But does such a dilemma really exist?

1.3.1 Recognizing Dualisms and Responding to Dilemmas

Dualisms come into play when two different, beneficial, and mutually exclusive classes of action are recognized in our socially constructed world. Dualisms can fade – or at least their distinctive boundaries can be softened and made permeable – when the world is seen as a holistic system, not as an arena in which dichotomous either/or decisions and calculated tradeoffs have to be made. CSR generally assumes as holistic world-view. Indeed, the urgency and necessity of CSR thinking is predicated on assumptions of the inter-connectedness, inter-dependency, and inter-relatedness of the social, economic, and natural worlds [28].

Significantly, it is not the corporate as an *entity* that has to make these decisions about behavior. Rather, these will be made by the corporation's strategic leaders and owners (nominally its shareholders) through the corporate governance structure. Corporations are familiar with dualisms, just as they are familiar with the dichotomies and the decision-making that accompany them. They are constantly faced with alternative courses of action that seem mutually exclusive [29, 30]. Should they enter into periods of disruptive growth or of relative stability? Should they embrace creative innovation or focus on conservative consolidation? Should they continue to rely on the present exploitation of resources or embark on an active exploration of new resources? Corporations might also see the same dualisms, options, and tradeoffs when confronted with CSR. Should they give greater weight – not necessarily preference – to socially responsible outcomes or prioritize and address economic profit and shareholder enrichment?

Corporations constitute a single class of legal entities, but they have different internalized value systems, missions, and objectives. Some might be governmental units or quasi-governmental agencies, constituting what is often identified as the public ("second") sector. Some might operate as not-for-profit enterprises and NGOs in the civic ("third") sector. Increasingly, some organizations are created to deal with social and community concerns – the social enterprises and blended value organization of the "fourth" sector. Organizations in these three sectors are mindful of CSR and will undoubtedly prioritize socially responsive and responsible behavior. Although they probably have a corporate form, they will pursue socially engagement using that legal form as a defensive *shield*, not as an aggressive *sword*. Indeed, most states in the United States allow social responsibility clauses to be inserted in the organization's articles of incorporation [31, 32].

However, in the dominant *private sector* (the "first" sector), most corporate managers and leaders sense an inherent duality in the very notion of CSR – *corporate* infers a specific cluster of assumptions and priorities; *social responsivity* is quite a different and oppositional notion. For many CSR, as a construct, is either an oxymoron or an irreconcilable paradox. The rules of the game for the corporation seem to cluster around preserving the entity, preferentially favoring insiders (shareholders), and optimizing wealth accumulation in economic contexts. These rules seem antithetical, or at least incompatible, with any notion of social responsibility or any sense of regard for the broader social worlds within which the corporation operates.

The corporation, when confronted with external demands for CSR, might then seem to be at an *impasse* rather than facing a *dilemma*. An impasse is an inability to decide because neither option is attractive or acceptable; a dilemma is when a choice has to be made between two equally beneficial possibilities but mutually exclusive options.

1.3.2 Reframing Dualisms, Avoiding Dilemmas, and Negotiating Impasses

Over the last 50 years, CSR has been extolled as an inescapable logic or rejected as an impossible demand. Over the same period, but not necessarily synchronized with these ebbs and flows, the corporate world has seen itself as either embattled or empowered. In the US corporate world, it has been argued that the relatively muted and accommodating nature of present-day CSR can be traced back to "the seismic changes which took place in the corporate world during the neoliberal counter-revolution of the 1980s and 1990s" when it became an *"adjunct* to the revived and reinvigorated, shareholder-oriented conception of the corporation, which has an appeal to both corporations and their critics" [33, p. 78, emphasis in original].

Corporate social responsibility became an adjunct issue – rather than a central one – when corporations realized that the apparently counter-intuitive behaviors required by CSR could actually be incorporated in successful and strategical ways into the new rules of an old game. It was recognized that the broadly social responsible versus the narrowly economical responsible dualism could be reframed in ways that did not force either/or decisions. It could be reframed in ways that required neither a corporate acceptance of the underlying values of CSR nor a corporate rejection that would undoubtedly have led to social condemnation or possible consumer retaliation.

It made better sense to recognize CSR as an issue of *economics* rather than as an issue of *social responsibility*. It made even better sense to adopt a commitment to CSR principles, at least rhetorically, as a way of preventing governmental interference in the markets, moving toward deregulating existing consumer markets, and projecting a favorable and beneficent corporate image. From a strategic perspective, "CSR has become a key element in corporate strategies to stave off direct government regulation and public criticism by projecting an image of corporate responsibility and fairness in a world where inequality and social injustice are growing" [33, p. 90].

By reframing apparent dualisms and avoiding the dilemma of choice, corporations are able to promote their version of a modified and ameliorative CSR, while at the same time continuing to provide preferentially advantages and increased wealth-accumulation benefits for their shareholders. This certainly seems to be the case with most of the major corporations (those listed in the Global Fortune 250) in North America and Western Europe, where recent figures show that almost 95% have embraced and actively report on their CSR initiatives [34].

1.3.3 Internal Stakeholders and Micro-level CSR Dilemmas

Potential dilemmas exist not only for the corporate as an entity but also for those who work within them. In recent years, there has been considerable interest in the challenges and opportunities provided by CSR at the micro-level, that is, at the level of: (a) the individual organizational member, as opposed to the corporate level and (b) the personal and the psychological, as opposed to the collective and the sociological. These micro-level interests tend to center on how current organizational participants personally respond to CSR initiatives, or their absence, and the impact of these personal responses on the individual's relationship with the organization – as measured by employee retention rates, institutional loyalty, and perceived corporate attractiveness in employee recruitment efforts and subsequent hiring patterns [35–37].

Corporate social responsibility can be defined as:

> caring for the well-being of others and the environment with the purpose of also creating value for the business. … manifested in the strategies and operating practices that a company develops in operationalizing its relationships with and impacts on the well-being of all of its key stakeholders and the natural environment [38, p. 171].

In this definition, organizational members see themselves, as key stakeholders, being included in the corporate CSR mission. Employees and managers, who themselves place value on the well-being of others in society, may find it easier to self-define and self-identify as organizational members of companies that are dedicated to CSR principles. They may find themselves in a situation of contributing efforts and creativity to organizations involved with positive aspects of CSR on which they, as individuals, place value. In other words, "individual with integrated CSR values will not promote CSR because they feel pressure to do so. … [instead] they will act to embody CSR values because the values have been internalized and integrated into their self-concept" [39, p. 76].

Organizational participants, particularly those at management levels, who are personally disposed to CSR approaches and who are engaged in a corporation with a similar outlook may: (a) be able and willing to contribute their *whole being* to corporate efforts and engagement, unrestrained by any personal conflict or concerns about CSR deficiencies; (b) respond favorably to corporate efforts because they detect an element of *self-interest*, since they too are significant stakeholders in the enterprise; (c) have a greater degree of ethical satisfaction and *moral appreciation* in their dealings and engagement with the corporation; and (d) gain an increased *sense of self-esteem* and social standing outside the corporation through their association with an organization that reflects their own values [40].

Current research at the micro-level suggests that corporate adoption of CSR provides both employees and managers, as significant stakeholders, and their companies with a number of benefits: positive organizational identity, reduced turnover and intent to leave, and increased social capital in the community for workers associated with the company. Adopting and adhering to CSR goals and values may also provide the firm with advantages in recruitment and hiring candidates who identify and support these values. Where a company embraces CSR approaches, organizational members have no dilemma, personal concern, or conflicted self-identification; instead, they can feel enthusiastic with their contributions to the firm's activities.

If, however, the organization has only a tepid or rhetorical interest in CSR engagement – reflected in a dissonance between public image and internal processes, procedures, policies, and reward systems – employees and managers may find themselves conflicted. They might find their continuing organizational efforts problematic, leading to increased turnover or intent to leave [36–38]. Naturally, these negative outcomes need to be interpreted contextually: What is the strength of the individual's commitment to CSR? What are sector employment and unemployment rates? How easy is employment mobility within that sector? Employee disengagement might materialize, creative involvement might decrease, but employees might not have the opportunity or inclination to translate this negativity into decisive action.

1.4 Triple Bottom Lines and Trilemmas

The definition of CSR has shifted to reflect different issues and areas of problematic concern [41]. At the beginning of the 1980s, Carroll suggested that socially responsible organizations needed to be actively concerned with the legitimate "economic, legal, ethical, and discretionary (philanthropic) expectations that society has of organizations at a given point in time" [42, p. 499]. By the early 1990s, the list of focal areas had expanded to provide more detail, with Wood understanding that the socially responsible organization should be concerned with the "business organization's configuration of principles of social responsibility, processes of social responsiveness, and policies, programs, and observable outcomes as they relate to the firm's societal relationships" [43, p. 693].

Early in the 2000s, Sandra Waddock reviewed the burgeoning and nuance terminology – corporate citizenship, CSR, corporate responsibility, corporate social performance, and business citizenship – that has fractured the subject. She considered that broadest notion was *corporate responsibility*, that is, "the degree of (ir)responsibility manifested in a company's strategies and operating practices as they impact stakeholders and the natural environment day to day" and noted that "this terminology is increasingly being used in business practice as a substitute or alternative for corporate citizenship" [44, p. 10]. Having defined corporate responsibility generally, she then argued that *CSR* represents a subset of concerns that "deals with a company's voluntary/discretionary relationships with its societal and community stakeholders.... frequently operationalized as community relations, philanthropic, multisector collaboration, or volunteer activities" [44, p. 10].

More recently, CSR definitions have been operationalized to include measurement criteria. For example, Aguinis defines CSR – which he prefers to term *organizational responsibility* – as the "context-specific organizational actions and policies that take into account stakeholders' expectations and the TBL of economic, social, and environmental performance" [45, p. 858]. He argues that this definition addresses four related issues: (a) it refers to *all organization*, not simply to for-profit corporations; (b) it stresses a *proactive intent* of "doing the right thing," not simply a passive concern for "doing no harm"; (c) it focuses on stakeholder expectations inferring the prioritization of *positive outcomes*, not simply the mitigation of negative or harmful ones; and (d) it includes *all stakeholders*, not just to a preferential class of shareholders [45].

Operationally, this definition purposefully leaves *stakeholder expectations* vague and contentious and provides no means by which these expectations might be either identified or measured. However, it does introduce an active commitment to CSR outcomes through reference to a *TBL* formula, which attempts to quantify and render transparent the organization's performance in economic, social, and environmental terms.

1.4.1 Triple Bottom Line Perspectives

The TBL perspective began in the early 1990s when John Elkington introduced it in an article, subsequently expanded into a book, dealing with ecological sustainability and the sustainable corporation [46, 47]. Elkington's premise was to shift corporate thinking from resource consumption to resource availability and sustainability. Rather than relying on the traditional accounting perspective that prioritized firm success based on a single economic metric – *net income*, often referred to as "the bottom line" – firms were encouraged to evaluate their performance on two other dimensions that considered social responsiveness and environmental responsibility. Because of the nature of the outcomes being evaluated, the TBL approach has also been referred to as the *3P Approach* (profit, people, and planet). Over the following 20 years, TBL has become a significant measure of the social and environmental responsibility for large-scale corporations that have the resources to evaluate these dimensions and the transparency to communicate the results [34].

A number of empirical studies and meta-analyses seem to indicate that TBL is a useful way of highlighting the firm's sustainability activities and economic performance. More controversially, some studies have suggested a positive relationship between these two variables [48, 49]. There are, however, significant problems associated with identifying and measuring these variables – particularly the corporation's sustainability activities – and a number of scholars have challenged the results of these studies, suggesting that there is little evidence for a positive relationship between sustainability efforts and economic performance and questioning the ability to produce credible or reliable data [50, 51].

Much of the impetus for TBL came from sustainability concerns rather than from a social responsibility interests but these concerns were often couched in qualitative, rather than quantitative, terms. The appropriate, valid, and reliable measurement of both sustainability and social responsibility dimensions has proved particularly problematic and is an area of ongoing research, even though to date it has yielded few significant results [52–54]. After reviewing the TBL literature, Norman and MacDonald concluded that more meaningful principles were required to inform and guide research and that "vague and literally meaningless principles like those implied by the TBL are best only for facilitating hypocrisy" [52, p. 257]. Needless to say, these conclusions have been robustly refuted by TBL proponents [55].

1.4.2 Corporate Trilemmas and Sensemaking

Triple bottom line initiatives, advocated by those who prioritize social responsibility and adopted by firms that place greater emphasis on financial and economic results, would seem to confront the corporation with a trilemma. Trilemmas occur when three equally beneficial outcomes are envisaged but selecting any two of them automatically means that the third is rendered unattainable. True trilemmas occur in a number of areas of economics, the best known of which is the "impossible trinity" that confronts a country when it seeks to simultaneously achieve monetary sovereignty, fixed exchange rate stability, and capital market mobility. These three objectives are interrelated but a decision to select any two means that the third is not possible. Faced with such a trilemma, the nation state must decide which of the two outcomes are more beneficial and accept that by pursuing these – for example, independent monetary policy and fix exchange rate stability – the third will become unattainable [56–58].

However, with TBL initiatives, it is unlikely that the economically centered firm will be placed in a trilemma situation because the firm will not recognize that economic, social, and sustainable bottom line optimizations exist as independent choices. Within the economic rules of the corporate game: (a) there will undoubtedly be an appreciation of the primacy of economic and financial outcomes and (b) social and sustainable outcomes will be recognized but subsumed into the economic calculus. The game will not be to select two optimal goals from the impossible trinity because the trinity is not recognized as true set of alternatives in a neoliberal economic world. Instead, the firm will recognize them as independent possibilities, monetarizes them, and sees them in terms of economic advantage such as increased reputational value or enhanced consumer approval.

From a strategic perspective – and situated within an exclusively economic arena – the firm will most likely consider non-economic responses and "determine if an increase in consumer demand or a willingness of consumers to pay a price differential (premium) for the socially responsible product/service will cover the differential cost" [59, p. 203]. Indeed, research seems to suggest that in TBL considerations firms do pay more attention to people issues (social) than to planet issues (sustainability) because: (a) consumer responsibility initiatives are easier to accomplish and involve lower economic income tradeoffs and (b) even "low-effort behaviors" in sustainability have the potential to "come across as greenwashing – the overuse of sustainability-centered vocabulary for the purpose of marketing or public relations" [60, p. 19].

Rather than entering social responsibility or green arenas and playing the games there, the corporation is likely to have a strategical preference for representing non-economic concerns in economic ways. Diligent box-ticking on the TBL statement, reassuring declarations of a commitment to social responsibility, and fortuitous greenwashing provide economic advantage because they elicit positive responses from consumers and other stakeholders [61, 62].

The attempt is not to mislead or to misdirect. Misdirection would suggest that multiple options were possible and that the corporation is diverting attention from those possibilities. From the corporation's perspective, there was only one option – an economic and financial one. In communicating its decisions and behaviors, the corporation may de-emphasize the economic calculus and portray its actions in a manner that resonates more convincingly with the expectations of its relevant stakeholders [63, 64].

The trilemma of economics, social responsibility, and ecological sustainability is not confronted and impossible choices are never made because only one alternative – financial performance and related shareholder wealth maximization – is considered, not three. Trilemmas do not arise when there is only one beneficial alternative to be selected. Nevertheless, TBL perspectives and accounting do provide a useful talking point because they suggest alternatives for corporate behavior, even although they might not necessarily promote it. In a recent study of organizations that use the TBL approach, Glavas and Mish suggest that "although we are not arguing the purpose of the firm, nor are we even suggesting that more than a minority of firms will embrace TBL models, we do argue that TBL firms demonstrate that an alternative model is in operation" [65, p. 640].

1.5 The Boundaries of Corporate Social Responsibility

The central issue for CSR is not that it is concerned with a futile battle of world-views, but that it involves the humility and essential wisdom to consider alternatives world-views. In making sense of the world in which they operate, corporations may believe they have limited choices about the rules of the game. Indeed, they may believe that they have no choices at all in CSR because it does not belong to that world and is not amenable to corporate sensemaking unless it can be monetarized, commodified, or in some other way converted into an economic good [66, 67].

However, there are different worlds and there are different choices, even if they are not presently recognized. The *corporate entity* will not engage – as an abstraction or legal fiction – in a reconsidered world, in new sensemaking, or in making new choices. Instead, it will be those *individuals* who populate the corporation that do so, and it might be that "through building the capacity and capability on this individual level that the organization as a whole acquires understanding of the implications and thus the local (collective) meaning of CSR" [68, p. 221].

1.5.1 Motivations for Corporate Social Responsibility

Although the corporate form has its ancient origins in socially responsive undertakings, it provides a particularly well-designed instrument for profit-seeking and wealth accumulation. It is protected from many of the risk assumptions of other non-corporate market players and has been provided with asymmetrical advantages in its dealings with the economic and social worlds. The question is why, given these differential characteristics and its non-human status, the corporation should consider CSR perspectives. There are three generally given reasons for such behavior [2, 8, 68, 69]:

- *Intrinsic motivation:* CSR might be adopted intrinsic reasons, in which the organization is positively motivated by a concern for those who are impacted by it. The core sentiment might rest on the organization's entrenched cultural values, on its internalized sense of ethical duty and obligation, or on genuine altruism. Some, or all, of these elements might be present within corporate entities that are inherently motivated by profit-seeking imperatives. However, it seems likely that these core values will be moderated and suppressed by an economic calculus, recognized roles of agency and fiduciary duty toward shareholders, and concerns about the adoption of CSR behavior on short-term profit and long-term wealth accumulation.
- *Extrinsic motivation:* Here, the focus is not on those who are outside the organization, but rather for those inside it – particularly its senior management, governance mechanism, and shareholders. In viewing potential corporate responsiveness and responsibility, these organizations are likely to be motivated by economic considerations and the extent to which CSR can be incorporated into that economic domain and provide a comparative advantage in the marketplace. Organizations motivated by extrinsic factors are likely to consider both external factors, such as the responses of consumers and other relevant stakeholders, and internal factors, such as organizational gains resulting from possible increases in employee performance, commitment, and loyalty [38, 70]. Although extrinsically motivated organizations may be concerned with enhanced economic performance, the likely rhetoric and sentiment within the company may well reject this. Instead, it will project more concerned and altruistic motives – feelings that will undoubtedly be encouraged through the organization's public relations apparatus and probably be genuinely shared by a majority of stakeholders [61, 62, 71, 72].

- *Stakeholder motivation:* Here, a shift to CSR is usually initiated and sustained by the concern of those stakeholders who are able to exert influence on the organization. Shareholders, working through the corporation's governance mechanism, may feel that the firm's existing CSR does not align with their own interests, concerns, or sentiments. They may be motivated by economic considerations (market competitiveness and brand image), by ethical and socially responsive philosophies, or simply because they possess a different world-view in which the restricted and economically limited instrumentality of the corporation seems too disconnected from their personally experienced world concerns. There is evidence that increasingly stakeholders, particularly shareholders, believe that corporations should play a more responsible role in society – or at least a less irresponsible role – and that it is right for those who possess the power and leverage to steer corporate performance in such a direction [73, 74].

Each of these motivations represents a pathway to CSR, but for the for-profit corporation, it is more likely that the propelling forces will come either from an adaption to the extrinsic values that surround the corporation or from stakeholder pressure – and that those stakeholders, given their power differentials, are likely to be shareholders. Additionally, corporate moves toward CSR are more likely to be proactive rather than reactive, since there is a general understanding that many stakeholders and publics relate negatively to reactive (forced) CSR responses [75, 76].

1.5.2 Parallel Universes and Porous Boundaries

The literature of CSR is extensive, confusing, and often contradictory. For some, particularly those outside the corporate world, CSR is a narrative that has a compelling message and an irrefutable logic. They predominantly rely on considerations of justice, ethics, and equity – not to say of humility, compassion, and mutual responsibility. In their minds, the game is about a collective wellbeing, a reality of interdependence, and an overarching concern fabric of society.

For others, particularly those located within the corporation, CSR is no more than an adjunct concept that does not authentically belong to the corporate world, but which might be adopted if it serves a good economic purpose and conforms to a profit-centered logic. They do not reject CSR values *per se*. Rather, they have another game in mind – a game with different players, different rules, and different goals. Their game is predominantly doctrinal and its rules involve neoliberal economic values, an unfettered vision of capitalism, and an often recited article of faith that good for the many emerges as the unintended consequence of the unbridled self-interest of the few [4, 7, 8]. There is no single CSR narrative. Instead, there are two – each held by a different set of actors, separated by different world-visions, and developed in different universes. Waddock has perceptively noted that the divergent academic-based and practitioner-based understandings of CSR "seem to have evolved in parallel, sometimes overlapping but sometimes universes apart" [44, p. 5].

It is easy to see why these two unconnected universes have been developed and just as easy to see why they have remained apart. From a managerial perspective – that is, from the interiority of the corporation – there are three key elements that give rise to a powerful, persistent, and pervasive world-view.

- **The power of socialization:** Corporate managers are socialized in the norms of profit-making and wealth-accumulation, albeit gradually and indirectly. They are immersed in environments that recognize ethical values but which require, measure, and reward economic behavior. Many quickly realize that they are operating in a market-driven universe that follows the celestial mechanics of neoliberal economics, the rule of profit-maximization, and the injunction to engage in "open and free competition without deception or fraud" [3, p. 133]. In this universe, there is only one game and losing is not a desired outcome. To the extent that CSR can be modified to conform to the rules of the game it is welcome, even although it remains something of an alien concept. These perspectives and values are not explicitly taught – they are learned and replicated. They constitute the cultural norms of the corporation and, as such, are internalized and adopted.

- **The specter of the corporation:** Socialization takes place through the social interaction and exchange between other individuals in the corporation. However, a striking aspect of this socialization is that it is shaped and molded the by non-human corporate entity itself. Within the otherness of the corporate sphere, the thinking, feeling, and behavior of human actors are shaped by its constructed form. The properties of the corporation – its perpetuated existence, asymmetrical asset protection, and preferential risk status – are not available or replicated in the social world. Thus, it is inevitable that those who populate it will sense that they have been set in a different world, with properties and capabilities that differ from their experiential world. It is not so much that human actors are changed by their corporate activities, but rather that what they can do and what they should do are circumscribed by the presence of the corporate veil.

- **The failings of managerial education:** Those who enter the corporation are often formally educated for that experience; however, there is much criticism of the business school for having failed to provide a pre-business experience that is sufficiently broad, critical, reflective, or ethically attuned. Bennis and O'Toole give business schools a failing grade for having lost their way and castigate them for their restricted vision, technocratic preoccupation, and dearth of experienced professors "who, collectively, hold a variety of skills and interests that cover territory as broad and as deep as business itself" [77, p. 104]. The business school's lack of responsiveness to issues of CSR has been highlighted by many others, some of whom question the continuing accreditation of these schools and their right to operate [78]. Others have forcefully objected to a lack of concern for ethics and responsibility, to the perpetuation and showcasing of "Enron-like" behaviors, and to a lack of leadership – all of which makes the business school and their accreditation agencies look like "silent partners in corporate crime" [79]. Yet others critics decry the present standard of responsibility management and urge business schools make this a more prominent and incisive part of their undergraduate and graduate teaching [80, 81].

It is easy to understand how a separate corporate universe can be created and perpetuated. It is easy to understand why the corporation is apart from the universe that lies beyond it and around it. And it is just as easy to appreciate that in its isolated parallel universe, the corporation recognizes and plays by a set of rules that are fundamentally different from the social world in which it operates. However, the consequences are significant, certainly for those who believe that they have a stake in corporate behavior and particularly those who deal directly with the corporation.

Consumers, most demonstrably in the finance and pharmaceutical sectors, are all too often viewed as an exploitable resource and the social impact of increasing consumer debt, egregious lending practices, and health insecurity are staggering and unsustainable [13]. Employees increasingly face extreme unpredictability and insecurity in their terms of employment. They are unilaterally forced into arrangements that systematically reduce their wages in real terms and into zero-hour contacts that allow corporations to punitively commodify labor, abandon traditional psychological contracts, and avoid any kind of employee benefits [82].

At the same time, environmental irresponsibility and retaliatory consumer boycotts are whitewashed through effective corporate communication strategies and selective media attention [83, 84]. US corporate profits have soared, staggering CEO remuneration and bonuses earned for lackluster managerial performance are indefensible, stock markets separated from the real underlying economy are at record highs, and wealth inequality is at an all-time high. Under the present US political regime, deregulation has increased to provide "level playing fields" for the favored few, safety and protection regulations have been set aside by well-funded corporate lobbies, and a fiercer more aggressive and combative capitalism – curiously grounded in trade restrictions, protectionism, and global isolationism – has been lauded.

In this parallel universe, the few are privileged at the expense of the many. The primacy of a harsh neoliberal economic doctrine is asserted to the exclusion of vague notions of social cohesion and social responsibility. Why, in these times and in these places, should corporations consider initiating or pursuing serious CSR actions, even to offset or obscure their past and present irresponsibility? [85, 86]

Of course, there is a glaring problem. Corporations do not inhabit an alternative universe, even though they may think they do. They can operate only in *our universe* that is populated by real people and which is filled with equally real social institution and communal concerns. The boundaries of the corporate world are fictions, albeit substantiated and preserved by those inside the corporate world. Corporate boundaries delimit behaviors that would otherwise be considered short-sighted, irresponsible, and predatory. Nevertheless, these boundaries are potentially *porous* and *permeable*. They separate the distinctiveness of the corporate universe from the social universe, but they also allow for the flow of ideas and understanding between them. CSR will only move from a sparingly applied cosmetic accommodation to an authentic engagement when the realities and consequences of that porosity are grasped.

1.6 Conclusion

Even at the time when they were first created, there must have been concern about the likely behavior of non-human corporations in human society and speculation on how corporations and natural persons would coexist. This is not to reify the corporate entity, seeing it as a totally independent actor driven by its own innate behavior. Corporations operate only in the social world when they are peopled by individuals but, in entering the corporation, it is suggested that individuals enter a different reality where different possibilities are available and different games can be played. This is not to argue that corporate members are dehumanized or in some way transformed into other-than-human creatures, but it is to suggest that the corporate form has a significant and altering impact on those who manage it and lead it. Within the corporation, new possibilities are presented and old responsibilities are dimmed, if not totally extinguished. It is important for corporate management and leaders to be aware of the power of the transformative influences that surround them and which may come to guide them.

Following the financial implosion of late 2007, and the continuing Great Recession that followed it, many would have thought that *corporate irresponsibility* was so blatant and destructive that it would inevitably be replaced by an ethos of doing no further harm – perhaps, even of beginning to do societal good. However, one of the remarkable realities of our worlds – both corporate and social – is that we have the ability to forget and the power not to remember. These can serve useful purposes for the individual, but they can also have negative results for organizations and for society generally because they "can facilitate an inability to learn from past mistakes, encouraging the repetition of questionable and even harmful courses of action ... even serve as a sort of resource for perpetuating questionable behavior in a systematic fashion" [87, p. 734].

There is certainly an increasing interest in CSR, but in all too many ways that interest is superficial and cynical. Despite the consequences of corporate irresponsibility, and beyond the era of the sociopathic masters of the universe, there is little evidence to suggest that corporations have refrained from what seem to most to be patent social harm or that they have learned from their misdeeds, if indeed they can come to regard these actions as misdeeds. For CSR to become a normative practice, more effort and thoughtfulness has to be spent on exploring the operational landscapes, value assumptions, and perceived constraints or organizations and of the world within which they operate.

Corporate social responsibility has a long and contested history, as a movement and as an area of disciplinary study, reaching back least half a century, even though social responsibility and irresponsibility have been recurring themes throughout a much longer history. In a recent review of almost 600 journal articles and a 100 books and chapters in edited books, Aguinis and Glavas consider what we know and what we do not know about the topic. In their conclusion, they suggest that CSR can be used "as a conduit to test management theories in the context of society ... [helping] us leave the world a better place than we found" [88, p. 960].

Many will find that prediction welcoming and reassuring. But that prediction cannot come to fruition until management theorists and management practitioners enter into a dialog that is mutually intelligible. They need to share their knowledge and their experience, and they have to move into each other's worlds. CSR is not a *battle* of rhetorical claims and assertions. Rather, it is an *impasse* that results from competing world-views, contextual behaviors, and the inability to enter – and more problematically to leave – the seemingly parallel universes that we have constructed.

To move beyond the impasse will require an appreciable shift in mindsets and in sensemaking. It will require a shared and concerted effort to re-evaluate our existing economic models and legal structures – re-evaluations that go well beyond the scope of managerial preferences. It will require a critical determination of the extent to which modern corporate behavior is the product of extractive capitalism and neoliberal economic doctrines. It will also require a re-radicalization of what constitutes social good and social well-being. Indeed, it is "difficult to escape the conclusion that without a re-radicalization ... CSR [will] disappoint its proponents [and] the corporate accountability movement will also achieve far less than its advocates hope" [33, p. 97].

References

[1] Lewis, D. L. (1976). *The Public Image of Henry Ford: An American Folk Hero and his Company*. Detroit MI: Wayne State University Press.

[2] de Jong, M. D. T., and van der Meer, M. (2015). How does it fit? Exploring the congruence between organizations and their corporate social responsibility (CSR) activities. *J. Bus. Ethics* 143, 71–83.

[3] Friedman, M. (2002). *Capitalism and Freedom*, 40th Edn. Chicago IL: University of Chicago Press.

[4] Okoye, A. (2009). Theorizing corporate social responsibility as an essentially contested concept: is a definition necessary? *J. Bus. Ethics* 89, 613–627.

[5] Scherer, A. G., and Palazzo, G. (2007). Toward a political conception of corporate responsibility: business and society seen from a Habermasian perspective. *Acad. Manag. Rev.* 32, 1096–1120.

[6] Dawkins, C. (2015). Agonistic pluralism and stakeholder engagement. *Bus. Ethics Q.* 25, 1–28.

[7] Friedman, A. L., and Miles, S. (2002). Developing stakeholder theory. *J. Manag. Stud.* 39, 1–21.

[8] Matten, D., and Moon, J. (2008). "Implicit" and "explicit" CSR: a conceptual framework for a comparative understanding of corporate social responsibility. *Acad. Manag. Rev.* 33, 404–424.

[9] Skilton, P. F., and Purdy, J. M. (2017). Authenticity, power, and pluralism: a framework for understanding stakeholder evaluations of corporate social responsibility activities. *Bus. Ethics Q.* 27, 99–123.

[10] Bowen, H. R. (1953). *Social Responsibilities of the Businessman.* New York, NY: Harper.

[11] Lange, D., and Washburn, N. T. (2012). Understanding attributions of corporate social irresponsibility. *Acad. Manag. Rev.* 37, 300–326.

[12] Murphy, P. E., and Schlegelmilch, B. B. (2013). Corporate social responsibility and corporate social irresponsibility: introduction to a special topic section. *J. Bus. Res.* 66, 1807–1813.

[13] Herzig, C., and Moon, J. (2013). Discourses on corporate social ir/responsibility in the financial sector. *J. Bus. Res.* 66, 1870–1880.

[14] Pies, I., Hielscher, S., and Beckmann, M. (2009). Moral commitments and the societal role of business: an ordonomic approach to corporate citizenship. *Bus. Ethics Q.* 19, 375–401.

[15] Kuran, T. (2005). The absence of the corporation in Islamic law: origins and persistence. *Am. J. Comp. Law* 53, 785–834.

[16] Pollock, F., and Maitland, F. W. (1898). *The History of English Law: Before the Time of Edward I,* Vol. 1, 2nd Edn. Cambridge: Cambridge University Press.

[17] Hansmann, H., Kraakman, R., and Squire, R. (2006). Law and the rise of the firm. *Harv. Law Rev.* 119, 1333–1403.

[18] Haldane, A. (2017). "Who owns a company? Speech by the executive director and chief economist of the bank of England," in *Proceedings of the University of Edinburgh Corporate Finance Conference*, Edinburgh.

[19] Rajan, R. G. (2012). Presidential address: the corporation in finance. *J. Finance*, 67, 1173–1217.

[20] Grantham, R. (2013). The corporate veil: an ingenious device. *Univ. Qld. Law J.* 32, 311–315.

[21] Cheng, T. K. (2017). The corporate veil doctrine revisited: a comparative study of the English and the U.S. corporate veil doctrines. *Boston Coll. Int. Comp. Law Rev.* 38, 329–412.

[22] Millon, D. K. (2006). *Piercing the Corporate Veil, Financial Responsibility, and the Limits of Limited Liability.* Lexington, VA: Washington and Lee University School of Law.

[23] Parker, D. (2015). The company in the 21st century: piercing the veil—reconceptualising the company under law. *J. Law Gov.* 10, 1–13.

[24] Schall, A. (2016). The new law of piercing the corporate veil in the UK. *Eur. Comp. Financ. Law Rev.* 13, 549–574.

[25] Ribstein, L. E. (2010). *The Rise of the Uncorporation.* New York, NY: Oxford University Press.

[26] Hayden, G. M., and Bodie, M. T. (2011). The uncorporation and the unraveling of "Nexus of Contracts" theory. *Michigan Law Rev.* 109, 1127–1144.

[27] Fischel, D. R. The corporate governance movement. *Vanderbilt Law Rev.* 35, 1259–1292.

[28] van Oosterhout, H., Pursey, P. M., and Heugens, A. R. (2008). "Much ado about nothing: a conceptual critique of corporate social responsibility," in *The Oxford Handbook of Corporate Social Responsibility,* eds A. Crane, D. Matten, A. McWilliams, J. Moon, and D. S. Siegel (Oxford: Oxford University Press), 197–223.

[29] Farjoun, M. (2010). Beyond dualism: stability and change as a duality. *Acad. Manag. Rev.* 35, 202–225.

[30] Sonenshein, S. (2016). Routines and creativity: from dualism to duality. *Organ. Sci.* 27, 739–758.

[31] Corry, O. (2010). "Defining and theorizing the third sector," in *Third Sector Research,* ed. R. Taylor (New York, NY: Springer Publishing), 11–20.

[32] Sabeti, H. *The Emerging fourth Sector.* Washington, DC: Aspen Institute.

[33] Ireland, P., and Pillay, R. G. (2010). "Corporate social responsibility in a neoliberal age," in *Corporate Social Responsibility and Regulatory Governance,* eds P. Utting and J. C. Marques (London: International Political Economy Series), 77–104.

[34] KPMG (2015). *The KPMG Survey of Corporate Responsibility Reporting 2015.* Available at: https://assets.kpmg.com/content/dam/kpmg/pdf/2016/02/kpmg-international-survey-of-corporate-responsibility-reporting-2015.pdf [accessed July 30, 2017].

[35] Jones, D. A., Willness, C. R., and Glavas, A. (2017). When corporate social responsibility (CSR) meets organizational psychology: new frontiers in micro-CSR research, and fulfilling a quid pro quo through multilevel insights. *Front. Psychol.* 8: 520.

[36] Gond, J.-P., El Akremi, A., Swaen, V., and Babu, N. (2017). The psychological microfoundations of corporate social responsibility: a person-centric systematic review. *J. Organ. Behav.* 38, 225–246.

[37] Rupp, D. E., and Mallory, D. B. (2015). Corporate social responsibility: psychological, person-centric, and progressing. *Annu. Rev. Organ. Psychol. Organ. Behav.* 2, 211–236.

[38] Glavas, A., and Kelley, K. (2014). The effects of perceived corporate social responsibility on employee attitudes. *Bus. Ethics Q.* 24, 165–202.

[39] Rupp, D. E., Williams, C. A., and Aguilera, R. V. (2010). "Increasing corporate social responsibility through stakeholder value international-ization (and the catalyzing effect of new governance): an application of organizational justice, self-determination, and social influence theories," in *Managerial Ethics: Managing the Psychology of Morality*, ed. M. Schminke New York, NY: Routledge, 69–88.

[40] Glavas, A. (2016). Corporate social responsibility and organizational psychology: an integrative review. *Front. Psychol.* 7:144.

[41] Carroll, A. B. (1999). Corporate social responsibility: evolution of a definitional construct. *Bus. Soc.* 38, 268–295.

[42] Carroll, A. B. (1979). A three-dimensional conceptual model of corporate performance. *Acad. Manag. Rev.* 4, 497–505.

[43] Wood, D. J. (1991). Corporate social performance revisited. *Acad. Manag. Rev.* 16, 691–718.

[44] Waddock, S. (2004). Parallel universes: companies, academics, and the progress of corporate citizenship. *Bus. Soc. Rev.* 109, 5–42.

[45] Aguinis, H. (2011). "Organizational responsibility: doing good and doing well," in *APA Handbook of Industrial and Organizational Psychology: Maintaining, Expanding, and Contracting the Organization*, Vol. 3, ed. S. Zedeck (Washington, DC: American Psychological Association), 855–879.

[46] Elkington, J. (1994). Towards the sustainable corporation: win-win-win business strategies for sustainable development. *Calif. Manag. Rev.* 36, 90–100.

[47] Elkington, J. (1997). *Cannibals with Forks: Triple Bottom Line of 21st Century Business*. Stoney Creek, CT: New Society Publishers.

[48] Orlitzky, M., Schmidt, F. L., and Rynes, S. L. (2003). Corporate social and financial performance: a meta-analysis. *Organ. Stud.* 24, 403–441.

[49] Peloza, J. (2009). The challenge of measuring financial impacts from investments in corporate social performance. *J. Manag.* 35, 1518–1541.

[50] Margolis, J. D., and Walsh, J. P. (2016). Misery loves companies: Rethinking social initiatives by business. *Adm. Sci. Q.* 48, 268–305.

[51] Wood, D. J. (2010). Measuring corporate social performance: a review. *Int. J. Manag. Rev.* 12, 50–84.

[52] Norman, W., and MacDonald, C. (2004). Getting to the bottom of the "triple bottom line". *Bus. Ethics Q.* 14, 243–262.

[53] Peloza, J., and Shang, J. (2011). How can corporate social responsibility activities create value for stakeholders? A systematic review. *J. Acad. Mark. Sci.* 39, 117–135.

[54] Hart, S. L., and Dowell, G. (2010). A natural-resource-based view of the firm: Fifteen years after. *J. Manag.* 37, 1464–1479.

[55] Pava, M. L. (2007). A response to "getting to the bottom of triple bottom line". *Bus. Ethics Q.* 17, 105–110.

[56] Aizenman, J., Chinn, M. D., and Ito, H. (2011). Surfing the waves of globalization: Asia and financial globalization in the context of the trilemma. *J. Jpn. Int. Econ.* 25, 290–320.

[57] Aizenman, J., Chinn, M. D., and Ito, H. (2010). The emerging global financial architecture: tracing and evaluating the new patterns of the trilemma's configurations. *J. Int. Money Finance* 29, 615–641.

[58] Mundell, R. A. (1963). Capital mobility and stabilization policy under fixed and flexible exchange rates. *Can. J. Econ. Political Sci.* 29, 475–485.

[59] McWilliams, A., Parhankangas, A., Coupet, J., Welch, E., and Barnum, D. T. (2016). Strategic decision making for the triple bottom line. *Bus. Strategy Environ.* 25, 193–204.

[60] Shnayder, L., van Rijnsoever, F. J., and Hekkert, M. P. (2015). Putting your money where your mouth is: why sustainability reporting based on the triple bottom line can be misleading. *PLOS ONE* 10:e0119036.

[61] Du, S., Bhattacharya, C. B., and Sen, S. (2007). Reaping relational rewards from corporate social responsibility: the role of competitive positioning. *Int. J. Res. Mark.* 24, 224–241.

[62] Du, S., Bhattacharya, C. B., and Sen, S. (2010). Maximizing business returns to corporate social responsibility (CSR): the role of CSR communication. *Int. J. Manag. Rev.* 12, 8–19.

[63] Crane, A., and Glozer, S. (2016). Researching corporate social responsibility communication: Themes, opportunities and challenges. *J. Manag. Stud.* 53, 1223–1252.

[64] Walter, B. L. (2014). "Corporate social responsibility communication: towards a phase model of strategic planning," in *Communicating Corporate Social Responsibility: Perspectives and Practice: Critical Studies on Corporate Responsibility, Governance and Sustainability*, Vol. 6, eds R. Tench, W. Sun, and B. Jones (Bingley: Emerald Group Publishing), 59–79.

[65] Glavas, A., and Mish, J. (2015). Resources and capabilities of triple bottom line firms: going over old or breaking new ground? *J. Bus. Ethics* 127, 623–642.

[66] Basu, K., and Palazzo, G. (2008). Corporate social responsibility: a process model of sensemaking. *Acad. Manag. Rev.* 33, 122–136.

[67] Edward Freeman, R., Wicks, A. C., and Parmar, B. (2004). Stakeholder theory and "the corporate objective revisited". *Organ. Sci.* 15, 364–369.

[68] Cramer, J., Jonker, J., and van der Heijden, A. (2004). Making sense of corporate social responsibility. *J. Bus. Ethics* 55, 215–222.

[69] Hemingway, C. A., and Maclagan. P. W. (2004). Managers' personal values as drivers of corporate social responsibility. *J. Bus. Ethics* 50, 33–44.

[70] Kim, Y. (2014). Strategic communication of corporate social responsibility (CSR): effects of stated motives and corporate reputation on stakeholder responses. *Public Relat. Rev.* 40, 838–840.

[71] Chernev, A., and Blair, S. (2015). Doing well by doing good: the benevolent halo of corporate social responsibility. *J. Consum. Res.* 41, 1412–1425.

[72] Malik, M. (2015). Value-enhancing capabilities of CSR: a brief review of contemporary literature. *J. Bus. Ethics* 127, 419–438.

[73] Becker-Olsen, K. L., Cudmore, B. A., and Hill, R. P. (2006). The impact of perceived corporate social responsibility on consumer behavior. *J. Bus. Res.* 59, 46–53.

[74] Morsing, M., and Schultz, M. (2006). Corporate social responsibility communication: stakeholder information, response and involvement strategies. *Bus. Ethics Eur. Rev.* 15, 323–338.

[75] Groza, M. D., Pronschinske, M. R., and Walker, M. (2011). Perceived organizational motives and consumer responses to proactive and reactive CSR. *J. Bus. Ethics* 102, 639–652.

[76] Chang, C.-H. (2015). Proactive and reactive corporate social responsibility: antecedent and consequence. *Manag. Decis.* 53, 451–468.

[77] Bennis, W. G., and O'Toole, J. (2005). How business schools lost their way. *Harv. Bus. Rev.* 83, 96–104.

[78] Murillo, D., and Vallentin, S. (2016). The business school's right to operate: responsibilization and resistance. *J. Bus. Ethics* 136, 743–757.

[79] Swanson, D. L., and Frederick, W. C. (2015). Are business schools silent partners in corporate crime? *J. Corp. Citizsh.* 9, 24–27.

[80] Nonet, G., Kassel, K., and Meijs, L. (2016). Understanding responsible management: emerging themes and variations from European business school programs. *J. Bus. Ethics* 139, 717–736.

[81] Hibbert, P., and Cunliffe, A. (2015). Responsible management: engaging moral reflexive practice through threshold concepts. *J. Bus. Ethics* 127, 177–188.

[82] Ingram, P., Yue, L. Q., and Rao, H. (2010). Trouble in store: probes, protests, and store openings by Wal-Mart, 1998–2007. *Am. J. Sociol.* 116, 53–92.

[83] Kleinnijenhuis, J., Schultz, F., Utz, S., and Oegema, D. (2015). The mediating role of the news in the BP oil spill crisis 2010: how US news is influenced by public relations and in turn influences public awareness, foreign news, and the share price. *Commun. Res.* 42, 408–428.

[84] McDonnell, M.-H., and King, B. (2013). Keeping up appearances: reputational threat and impression management after social movement boycotts. *Adm. Sci. Q.* 58, 387–419.

[85] Campbell, J. L. (2007). Why would corporations behave in socially responsible ways? An institutional theory of corporate social responsibility. *Acad. Manag. Rev.* 32, 946–967.

[86] Kotchen, M., and Moon, J. J. (2012). Corporate social responsibility for irresponsibility. *B.E. J. Econ. Anal. Policy* 12:55.

[87] Mena, S., Rintamäki, J., Fleming, P., and Spicer, A. (2016). On the forgetting of corporate irresponsibility. *Acad. Manag. Rev.* 41, 720–738.

[88] Aguinis, H., and Glavas, A. (2012). What we know and don't know about corporate social responsibility: a review and research agenda. *J. Manag.* 38, 932–968.

2

Future-Focused Entrepreneurship Assessment (FFEA)

Niko Roorda

Roorda Sustainability, Sprang-Capelle, The Netherlands.

Abstract

FFEA®, *Future-Focused Entrepreneurship Assessment*, is a system that enables companies and other organizations to investigate to which extent they are "fit for the future."

FFEA rhymes with "tea", is based on a management philosophy that describes four perspectives from which companies see themselves and their relations with the outside world. The perspectives, called "diligent," "targeted," "systemic," and "holistic," together form the basis for an assessment instrument that is the core of FFEA. The FFEA assessment is done by a group of stakeholders of the company (such as management, employees, customers, owners, suppliers, banks, neighbors, community representatives, societal or environmental action groups, etc.), who together reach consensus about (1) the current perspective of the company in relation to a series of different topics and (2) the "desired" perspective that is to be reached after, for instance, a couple of years.

The purpose is to investigate how likely it is that the company will survive, prosper, and fit within a sound society and environment. If weaknesses are found during the assessment, plans are made to make improvements, in order to increase the company's *"Future-proof Resilience."*

The conclusions of a FFEA assessment may be that only minor improvements are necessary. On the other hand, it may appear to be of vital interest to drastically change the way in which the company acts, thinks, or even sees itself, leading to a process called *"change the company mission."*

The result of the assessment is a concrete action plan.

The chapter first describes why this future-proof resilience of companies is directly linked to the future-proof resilience of society as a whole. The link between the two is CSR (corporate social responsibility). Motivations of companies for CSR are discussed, making clear why CSR is essential for their continuity. Cases are offered, such as Eastman Kodak, Xerox, the music industry, and the Royal Dutch Shell Company.

The four FFEA perspectives are introduced, after which the entire FFEA system is explained, including its assessment procedures. Next, it is shown how FFEA is to be applied as an integrated part of the quality management cycle. This is illustrated with two practical cases.

After a FFEA assessment is finished, it may be concluded that, as a part of the action plan, special attention is to be given to a number of FFEA's 30 topics. For most of these topics, FFEA offers extra tools, the so-called "FFEA Extensions." A selection of them is described in this chapter.

The final part of the chapter explains the origins and theoretical backgrounds of FFEA, enabling the reader to study the way in which the system was developed, based on a range of management theories.

One of the sources of FFEA is AISHE, which was developed – just like FFEA – by the author of the present chapter. AISHE, the "Assessment Instrument for Sustainability in Higher Education," was applied hundreds of times in higher education institutions, before it was used as the main source for FFEA.

The AISHE Certificate is currently applied as a model for a certificate that will be launched as an extension to FFEA. Companies which are awarded the FFEA Certificate can thus show that they are "Future-proof Resilient."

2.1 Motivations for CSR

2.1.1 Maslow for CSR

Corporate social responsibility is hot. Many companies practice it, ranging from multinational businesses to small- and medium-sized enterprises (SMEs). Their annual reports are silent witnesses of it, in various forms. Some companies present separate CSR Reports (e.g., Cisco [1]) or combined CSR and Sustainable Development Reports (e.g., Volkswagen [2]), next to a more general annual report. Others use integrated reporting (e.g., Denso [3]), in which CSR is accounted for as an essential aspect of all or most business activities. In some of such integrated reports, CSR as such is not mentioned explicitly but related terms are such as *social and environmental benefit, corporate government*, and/or *sustainable development* (e.g., Unilever [4]).

Why is CSR hot? For what reason have millions of companies all over the world adopted CSR as a bottom line? Actually, for a wide variety of reasons. In some cases, the main motivation is based on *fear*. Fear for all kinds of bad things that might happen, either to the company as a whole or to board members or managers, employees, stockholders, the local community, or the environment.

In other cases, the CSR mission and strategy is based on the opposite: *hope*. Hope to be able to contribute to a better world. To combat poverty, educate people, strengthen communities, save ecosystems, or simply to expand the company.

Where fear is the main motivator, relevant causes for that fear may be: the limited availability of energy; of raw materials; of legal maneuvering space; money; patents; and customers. Alternatively, fear may exist for legal liability or for reputation damage.

Hope, on the other hand, may be driven by idealism, but also by creative and innovative minds, market opportunities, or a thrive to create or increase prestige, respect, or even power, or simply the ambition to do business in a decent way.

Which kind of motivation is better? The answer to that question depends on the personal position and convictions of those who are judging. For some, CSR without idealism is without any real value, being at best an example of greenwashing or window-dressing, and at the worst a kind of hypocrisy. As an example, the McDonald's Corporation has been mentioned [5], as this company presents itself as a strong advocate of CSR, while at the same time being accused of abuses of labor rights and health safety standards, union-busting practices, and offering low wages. In cases like this, commercial interests seem to contradict idealistic motivations or ethical standards.

However, there is no real reason why commercial and idealistic motivations are in mutual conflict. This may be made clear with the aid of a simple scheme that is based on Maslow's famous motivations pyramid [6]. In this model, a hierarchy of needs is described, consisting of five layers. The most fundamental needs, according to Maslow, are presented at the lowest layers, as the left side of Figure 2.1 shows.

In Figure 2.1, these five layers of needs are compared with the needs of companies. [Wherever in this chapter the word "companies" is used, all kinds of organizations are implied, unless otherwise stated, including commercial companies and non-profit organizations such as schools, healthcare

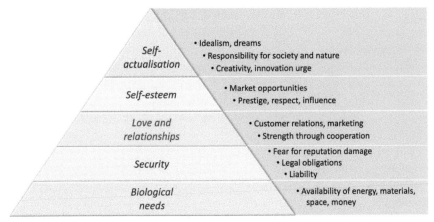

Figure 2.1   Maslow's hierarchy of needs [6] related to corporate motives for CSR.

institutions, government departments, non-governmental organization (NGOs), etc.] Actually, the comparison with Maslow's model of needs implies that these companies are conceptualized as a kind of living organism, for which their physical and mental needs act as motivations for their plans and activities [7].

Motivations based on *fears* are mainly present at the lowest two levels of the pyramid. There, the fear for a lack of energy, materials, money, or legal maneuvering space is to be found, as well as the fear for damages and liabilities. Both kinds of fear are presented as *biological* needs (also referred to as physiological needs) and the wish for *safety and security*.

The top three layers mainly represent hope, ranging from self-oriented hopes such as market opportunities to altruistic ones like dreams and ideals.

No doubt, all companies who dedicate attention to CSR do this based on not just one motive, but rather a combination of motives. Arguably, in many cases, such a *motivation mix* is related to all five levels of Maslow's pyramid, or at least to most of them. Consequently, when one zooms in on particular companies, the question should not be whether its CSR motives are purely idealistic or instead purely selfish; both extremes will almost certainly rarely be the case. On the one hand: 100% idealistic motives may be a threat to the company's financial position. Every company needs to make profit or at least provide sufficient income to continue its operations. Completely idealistic organizations may sooner or later fail by going bankrupt. On the other hand: 100% selfish motives for practicing CSR will carry the threat of being exposed, after which losing customers will also damage the company.

Later in this chapter (in Section 2.7), the *CSR Motivation Mix Model for Companies*, based on Maslow's model, will be presented as a management tool, along with other pragmatic tools that together are described as the *FFEA Extensions*, i.e., as accessories to the *FFEA* system which is the main topic of this chapter.

2.1.2 Future-Proof Resilience of Companies and Society

More and more companies and CSR organizations prove that idealistic and pragmatic motives don't conflict. To the contrary: analyses (e.g., [8]) show that in many cases, there is a strong positive correlation between ideals-based CSR and profit [9] even writes, linking CSR to sustainable development: "Sustainable front-runners are more successful than average businesses." This conclusion is supported by others, e.g., [10], who illustrates the reality of *corporate sustainable profitability* using a range of Swedish companies.

Table 2.1 illustrates how the interests of society and the environment, striving for a sound future, coincide with the same interests of companies, and how those companies can deploy CSR-based actions for the interest of both, contributing to what may be called *"Future-proof Resilience"* of both.

Table 2.1 makes clear that the continuity and the profitability of a company depend on its CSR strategy for many reasons. This is illustrated by the demise of many past companies, professions, and even entire business sectors that have ceased to exist for a range of reasons, many of which are present in Table 2.1.

In the 21st century, this is more true than ever before. It is interesting to combine the past, the present, and the future with the aid of a "Red List of Professions and Business Sectors," shaped after the IUCN Red List of Threatened Species [11]. Using the same terminology as the IUCN does for species of animals and plants, this Red List of Professions and Business Sectors looks like Table 2.2.

During discussions between the author of this chapter and members of company boards, frequently a question was asked to them:

- "Why do you expect that your company will still exist in, say, 20 years?"

It is striking that one kind of answer to this question has *never* been heard:

- "I am not at all sure that my company will exist in 20 years."

Table 2.1   CSR Actions for future-proof resilience

Level	Interests and Risks of Companies	CSR Actions
5. Actualization	● Dreams and ideals	● Evolve toward sustainable identity and mission.
	● Responsibility	● Enhance transparency and integrated reporting. Move from *shareholder value* to *stakeholder value.*
	● Creativity and innovation	● Stimulate creativity of personnel. Invest in research.
4. Esteem	● Market opportunities	● Offer innovative products/services. Open new markets. Be sensible to changes in society, technology, and economy.
	● Prestige, respect, and influence	● Be honest. Participate in societal projects. Don't discriminate. Offer fair wages.
3. Relations	● Continuous customer relations	● Trade fairly. Listen and respond to customers. Move from selling products to offering services.
	● Cooperation within the chain	● Adopt principles of cradle to cradle (C2C) and circular economy. Reject unsustainable suppliers.
	● Local cooperation	● Invest in relations with regional community and within business park.
2. Security	● Reputation, liability	● Offer safe and healthy products/ services and jobs. Recall products as soon as necessary. Offer generous indemnities.
	● Legal obligations	● Comply with laws and regulations. Comply with corporate codes and ethical standards.
1. Biological needs	● Climate change and energy scarcity	● Apply sustainable energy. Increase energy efficiency of production and products.
	● Nature damage and materials scarcity	● Reuse, recycle, and substitute. Design for disassembly.
	● Insufficient space	● Stimulate efficient, multifunctional use of space. Protect forests and ecosystems.
	● Solvability and liquidity	● No risky investments. Support sound banking system.

It is nice to see that entrepreneurs and managers believe in their company – although it must be said that in some cases a variation to this unheard answer was occasionally given:

● "Between now and 20 years, we may have been a partner in a merger."

Table 2.2 Examples of professions on the Red List of Professions and Business Sectors

Status	Examples
Extinct, locally extinct or critically endangered	Clog makers, telegraph operators, night soil collectors, fire makers, log drivers, cobblers, ice deliverers, lamp lighters, leech collectors, bowling pin setters, typesetters, elevator operators, resurrectionists, and knights.
Threatened	Chemical photo and film industry, millers, bridge keepers, travel agents, printing companies, desktop publishers, clock hand inspectors, paper mills, shepherds, CD music industry, and inner city shops.
Vulnerable	Bus drivers, typists, translators, livestock farmers, infantrymen, money printers, cashiers, librarians, software developers, financial analysts, gas stations, and oil companies.

A frequently given answer is:

- "We will still exist as a company, because we make such good products."

Or alternatively:

- "... we deliver such good services."

In spite of this answer, and related other ones, delivering good products or services is not a proof that the company will go on making it: there are many reasons, as Table 2.1 shows, why a company may die in spite of its excellent products. As society, science and technology change, so do customer preferences. The world is changing, and it does so faster and faster. Thus, it may well be that products with an excellent quality, that are highly popular at one moment, may lose customers' interests at the next. Alternatively, even if the products as such are still wanted, the company may get into trouble due to reputation damage. As a consequence, companies, professions, or entire business sectors may end gradually or abruptly.

2.2 (Not) Ready for the Future

One may discuss whether all entries in Table 2.2 are placed correctly. Probably, many more items are to be added. But the general idea is evident: many professions and sectors are threatened or have already disappeared. Even professions that perhaps 10 years ago looked solid and safe, such as drivers of buses, taxis, and trains, could die out in the next 10 or 20 years. In a report published in 2014 [12], CBRE Consulting warns that 50% of all occupations may have become obsolete within a few decades.

As a consequence, companies in many business sectors may go bankrupt on a large scale.

Many examples illustrate the truth of this fact. Take Eastman Kodak and Fujifilm, who thought around the year 2000 that chemical photography had a sound future. Here is the story of Kodak [7].

2.2.1 The Eastman Kodak Case

1888: The Kodak company is founded by George Eastman and Henry A. Strong.

1975: Eastman Kodak dominates the US market for photography, commanding 85% of camera sales and 90% of film sales.

Also 1975: Kodak develops the very first digital camera. But it is dropped, as the company fears it will threaten Kodak's own chemical photography business.

1994: Apple introduces the innovative QuickTake digital camera. Although it carries the Apple label, it is produced by Kodak, which proves that Kodak's digital knowhow is up to date.

2000: In the years around the start of the new millennium, Kodak does not develop a digital photography strategy. The company believes that its core business, traditional film, will not be threatened by the digital technology.

In the same years, Sony, Nikon, and Canon start flooding the market with digital cameras.

2004: While in the midst of a hasty U-turn toward digital technology, Eastman Kodak announces it will shed 15,000 jobs. A year later, the number is raised to 27,000. Its stock value plummets: see Figure 2.2.

2010: Kodak rapidly used up its cash reserves. Most of its revenues now come from patent licensing to competitors.

2012: Eastman Kodak is declared bankrupt. In an attempt for a restart, its photographic film, commercial scanners, and kiosk divisions are sold.

2013: Next, many of its patents are sold as well, for more than half a billion dollars. In August, the company emerges from bankruptcy after also abandoning personalized imaging and document imaging. In October, it returns to the New York Stock Exchange (NYSE), as Figure 2.2 shows.

In the 21st century, hardly anyone still buys old-fashioned rolls of film. Kodak started to move too late, years after young computer companies had already conquered the digital photography field, leaving no niche for Eastman's company.

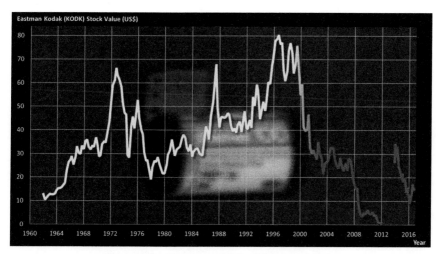

Figure 2.2 The rise and fall of Eastman Kodak.
Sources: [13,14]. Background photo: [15].

Kodak was *not* ready for the future.

It was not only Eastman Kodak that fell into the time trap. Its one-time main competitor, Fujifilm, made the same mistake and got into trouble too. Another company fared completely different: Xerox.

2.2.2 The Xerox Case

In 1938, the very first photocopy was generated in a New York laboratory. The first photocopier appeared in 1949. The process, called "xerography," was owned by a company that called itself the Xerox Corporation from 1961 onward, describing itself as "The Copying Company."

But the company did not stop there. Beginning in 1970, Xerox also investigated in computer technology. As computers were starting to conquer the world, Xerox realized at an early stage that there would come a day when computer monitors and hard drives would largely replace the role of paper documents. If that were to happen, the use of photocopiers would decline and this would ultimately – maybe only after decades had passed – spell the end for the Xerox Corporation.

This perspective of the future resulted in the company adopting a drastic change of course in the 1990s, when it launched a new image and officially called itself "The Document Company." On top of photocopiers, the company placed a larger focus on products and services for digital information and documentation.

Figure 2.3   Xerox: ready for the future. Kodak and Fuji: not ready.

Xerox was ready for the future; see Figure 2.3.

The Xerox case shows that it is essential for each and every company – whether commercial or non-profit – to investigate every few years whether it has a sound future. Threat to the company continuity may come from many sides. Perhaps, new technologies of competitors may make the present products obsolete. Perhaps the company makes use of child labor in far-away countries or pollutes the environment, causing customers to flee (voting with their feet) or governments to impose laws and regulations the company is unable to afford financially or mentally. A lack of CSR can be lethal.

If it seems that there is a serious threat to the company continuity, it is vital to investigate the reason why the company exists – the raison d'être or the mission – and see if it can be redefined, just like Xerox did. *Redefining the Company Mission*, as this exercise is called, is something that every company should do every now and then. Another example illustrates this principle again.

2.2.3 The Music Industry Case

LPs and CDs: once they were a cash cow for the music industry. But then, online file sharing started. Through Usenet and Napster, followed by Rapidshare, Megaupload, and torrent sites like Demonoid and ThePirateBay, millions of songs and compositions were downloaded for free – legally or otherwise. Soon, the music industry and the record stores were in big trouble. Sales and stocks plummeted, and the music companies complained loudly. Lawsuits were fought.

In an attempt to turn the tide, the music industry introduced paid downloads, together with digital rights management (DRM) that were supposed to block free downloads. In spite of these efforts, the revenues of recorded music sales continued to decrease, while paid downloads did not do well either.

The size of the loss is difficult to assess. According to the Institute for Policy Innovation (IPI), the effects were the following [16]:

> *"The US economy loses $12.5 billion in total output annually as a consequence of music theft. Sound recording piracy leads to the loss of 71,060 jobs to the US economy. Music theft also leads to the loss of $2.7 billion in earnings annually in both the sound recording industry and in downstream retail industries."*

Estimates of the loss, however, are heavily disputed [17], and are therefore hard to express graphically. Revenues of the music industry are much more objectively assessable, and are shown in Figure 2.4.

The music industry definitely had to reinvent itself. It was a classic case of the necessity of *redefining the company mission*, or rather: redefining the *sector* mission.

The tide turned, when streaming music was introduced. iTunes, Spotify, and others stormed the market and conquered a major market share. The effect is clearly visible in Figure 2.4, showing the expectations for the next decade.

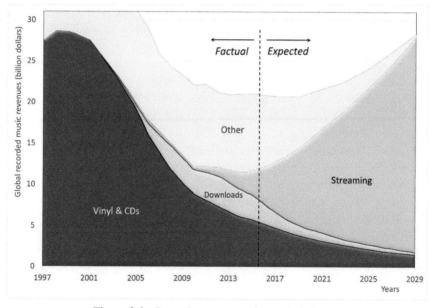

Figure 2.4 Streaming may save the music industry.

Sources: [18,19]. "Other": e.g., performance rights, film, and advertising tie-ins.

All of the above cases show that companies, professions, and business sectors are not invulnerable. They can be heavily damaged or even disappear. No sector can ever be future-proof.

In other words, a company that cares for its own future should also care for the future of society and the natural environment, and should adopt CSR and sustainable development as its bottom line.

Corporate social responsibility is where the future-proof resilience of a company and the future-proof resilience of society and the environment come together. In formula-shape:

> *Formula 1:*
> *CSR = Entrepreneuring from a perspective of the future of the company.*
>   = *Entrepreneuring from a perspective of the future of society*
>   *and the planet.*

2.3 The Four Perspectives of Future-Focused Entrepreneurship Assessment

"You probably spend a lot of time in cars, perhaps at the steering wheel. You make sure that your front window is clean and clear so that you can constantly see where you are heading. If this is not the case – perhaps you covered the windshield because of severe frost to prevent ice crystals – and you start driving while the cover is still on, you won't get very far. A hundred yards later, you will probably end up wrapped around a lamppost, or worse, you'll hit another car or a child crossing the street. Of course, nobody would do that. You would be a fool if you did: You don't go on the road blindly!

Have you ever realized that when riding in a car, you actually travel in two directions at once? First, when you drive on the road, you cover a distance expressed in miles or kilometers. Second, while you drive, you also move in time, and this is expressed in minutes or hours. Even if you are standing still, not adding any miles to your traveling distance, your temporal motion continues: hour upon hour, day by day, year after year. Why, then, do so many people travel blindly toward the future? (...)

It's always beautiful to see how an entrepreneur believes in his or her own product, but whether that is sufficient is doubtful. What if, between now and in twenty years, this excellent product becomes obsolete or is out-performed by something even better or cheaper? What if the people twenty years from now just don't need your product, excellent as it may be? Yes, entrepreneurs who don't think about such things travel blindly toward the future."

2.3.1 Traveling toward the Future

The above text is the introduction to the description of one of a set of competences for sustainable development ([20], Chapter 8). The entire set is called *RESFIA+D*, with the subtitle *The Seven Competences of the Sustainable Professional*. Whereas [20] offers a series of case descriptions to illustrate the model, which can be applied as an assessment and policy instrument by companies and individuals (see Section 2.9), the theoretical aspects are treated in [21], i.e., as a chapter in a book in the same series of books as the present one, with the same editors. Online details of RESFIA+D can be found in [22].

The opening text of this section is an introduction to competence *F: Future Orientation*. Competence *F* consists of three concrete achievements. The first is: *F1 – Think on different time scales: flexibly zoom in and out on short- and long-term approaches.*

Practically, this means:

- You zoom in: You analyze the opportunities and consequences of your work for the short term.
- Your short-term approach to a problem is aimed at tackling symptoms. (This is the *operational* approach.)
- You also zoom out: You investigate the options for the long term, if necessary even decades ahead, to improve your work fundamentally and in an innovative way.
- Where solving problems is the issue, you primarily aim at eradicating their causes. (This is the *strategic* approach.)
- This zooming in and out in time is done regularly and fluently, thanks to which you have continuous attention for both the short and the long term.

Redefining the company, mentioned earlier, is a typical example of the long-term, strategic approach. [20] goes on to explain this principle:

"Zooming in and out in time is easy to understand if you imagine the following situation: You are sitting quietly at home one night. Outside it is raining cats and dogs, but inside it feels nice and cozy until you suddenly realize that water is dripping from the ceiling. The roof is leaking! What do you do? Well, or course, the very first thing you do is get towels, wipe up the water, and place some buckets. The leaking roof is one thing, but you don't want your costly hardwood floor or broadloom carpet to get ruined as well."

Next, you have to find the cause. Something is wrong with the roof. Maybe some roof tiles have shifted, or perhaps a watertight strip is no longer watertight. In other words, you find the cause and repair the damage. You might even take another big step and replace the entire roof, or as a last resort, you may find another house and move. Wiping water up with towels and placing buckets – these are examples of short-term actions. They don't address the causes, but they prevent further damage. This is what acting operationally means.

"Repairing the roof represents a more profound approach. If you don't do it, you will still be emptying buckets next year. It is a tactical approach, aimed at a longer term than the towel operation. It's not a matter of minutes; rather, it takes days or weeks.

When you replace the roof or even move to another place altogether, then you apply a genuine long-term view. We call this a strategic approach, which involves considering many issues, not just acute problems. For example, you would ask yourself: How do I want to live in twenty years?"

Evidently, moving to another place is a metaphor for redefining the company mission.

The range from operational to strategic forms a frequently used series of different policy levels, which can be defined as follows:

An *operational* policy makes use of methods that can be applied immediately or in the short term, possibly (but not necessarily) based on tactical or strategic plans.

A *tactical* policy aims at an intermediate term, attempting to realize concrete targets of the organization or of a person, possibly (but not necessarily) derived from the strategic policy.

A *strategic* policy aims at the long term, endeavoring to realize fundamental goals based on the mission of an organization or on personal life goals.

Whereas this range of three policy levels is a classical one, used by many, this chapter adds a fourth policy level, which is essential to the *FFEA* system that is the main topic of the present chapter:

A *panoramic* policy aims at the long term of not only the most expected future, but based on a vision about a range of possible or thinkable futures, even actively participating in creating or supporting developments (gates) toward sustainable, societally preferred futures (utopias), and discourages developments toward less preferred and unsustainable ones (dystopias).

2.3.2 Company Perspectives

Just like human beings, companies act from a certain *perspective*: a view on itself, on the outside world, and on its position within that world. Due to this perspective, the company has expectations and ambitions about what it can do and should do. The perspective may have been made explicit in a company mission, describing the reason why the company exists and what it strives for. But the perspective may even go beyond such a mission, if it also defines the kind of society in which the company wishes to exist, and the notions about the range of possibilities the company possesses to give shape to this: its *action perspective*.

For many companies, their perspective is primarily related to one of the four policy levels described above. Although not all of their actions will be based entirely on just one policy level, assessments made with FFEA show that, when a wide range of topics are assessed, many of them are based mostly on one policy level.

It is the task of a FFEA assessment to discover the company perspective, and to enable decisions about adaptations of this perspective, thus raising the future-proof resilience of the company. This will be explained in detail in the next sections.

In order to make such an assessment possible, it is first essential to have a scale along which the various perspectives can be expressed. This is what Table 2.3 offers. The four perspectives, applied to companies, have all been given a name: *diligent, targeted, systemic, and holistic.*

The perspectives are numbered using the Greek character Π *(capital "pi"): $\Pi 1$–$\Pi 4$.*

As an example: a *targeted* company will, in many respects, act from an intermediate term, based primarily on tactical considerations, although in some respects, the other three perspectives may have a role in this company, too.

By the way: how long is a short, intermediate, or long term? It is impossible to define this in an absolute sense; it depends on the context. For businesses or communities, in most cases, a good general guideline is:

Short = now or in less than 6 months to 2 years
Intermediate = 3–5 years
Long = up to 20 or a 100 years (or more)

Table 2.3 Overview of the four perspectives

Perspective	Π1: Diligent	Π2: Targeted	Π3: Systemic	Π4: Holistic
Policy	Short term: *Operational*	Intermediate term: *Tactical*	Long term: *Strategic*	Panoramic: *Visionary*
General description	The organization primarily focuses on activities that are being carried out now or shortly, and spends little time on external trends that may affect the organization. It relies on current products or services, there are not many initiatives for innovation, and as far as they are, they will remain isolated and they will not or hardly be invested in. "Do not talk but work on," could be the motto.	The organization aims at fulfilling measurable goals of the established policy. For these purposes, new initiatives are being invested, which lead to visible results, e.g., in the form of improved or new products or services or new markets. The extent to which they help making the organization more future-proof has not been clearly investigated, as the organization does not pay much attention to long-term developments in society, science and technology, or in its own sector. The mission and core values of the organization are not really discussed with an open mind. As far as they are discussed, only a select part of the organization participates.	The organization knows who she is and who she wants to be. There is a clear and continuous alignment between short- and long-term developments in the outside world, the organization's mission, core values, activities, and future plans of the organization. All departments and ranks of the organizations are actively involved, bear responsibility, and are taken seriously by management. The organization operates as an organic whole.	The organization actively participates in society and its developments. It has an explicit vision of its position within society and the natural environment. It takes a visionary position about a range of possible or thinkable futures. It actively participates in creating or supporting developments (gates) toward sustainable, societally preferred futures (utopias); and discourages developments toward less preferred, unsustainable ones (dystopias).

Many exceptions to this rule exist. For example, if the roof is leaking, repairs (= intermediate to long term) occur within a matter of days or weeks at the most – hopefully.

2.3.3 Cumulative Perspectives

The range *Π1: diligent*, *Π2: targeted*, *Π3: systemic*, and *Π4: holistic* forms an ordinal scale. That is to say: there is a natural order in which they are sorted: from short to long, from narrow to wide. The holistic perspective obviously offers the widest view, both in time and in space, as well as in the level of creativity.

The descriptions of the consecutive perspectives are stacking, or with a more formal word: cumulative. This means that the descriptions for the first perspective (*Π1*) – at least insofar as they are positive and not restrictive – are still valid for perspective *Π2* and above, even though they are not again mentioned there. In the same way, the characteristics of perspective *Π2* are again applicable to perspective *Π3* and above, and so forth.

The principle is shown in Table 2.4.

As an example: where *Π2* (*targeted*) is characterized in part by "The organization aims at fulfilling measurable goals of the established policy," this will probably still be true for *Π3* (*systemic*) and *Π4* (*holistic*), fully or at least partially, although this not again mentioned in their columns in Table 2.3.

On the other hand: where another aspect of *Π2* is characterized by "The extent to which they help making the organization more future-proof has not been clearly investigated," this description is not positive but restrictive, and so it does *not* apply to *Π3* and *Π4*; *Π3* even explicitly claims otherwise.

Whenever FFEA is applied as an assessment instrument, as will be described soon, the four perspectives will always have to be interpreted in accordance with this cumulative principle.

Table 2.4   The cumulative nature of the four perspectives

Perspective	Π1: Diligent	Π2: Targeted	Π3: Systemic	Π4: Holistic
Characteristics	Descriptions of *Π1*	Descriptions of *Π1* (not mentioned again) + Descriptions of *Π2*	Descriptions of *Π1* (not mentioned again) + Descriptions of *Π2* (not mentioned again) + Descriptions of *Π3*	Descriptions of *Π1* (not mentioned again) + Descriptions of *Π2* (not mentioned again) + Descriptions of *Π3* (not mentioned again) + Descriptions of *Π4*

Whereas Table 2.3 offers general descriptions of the four perspectives, Table 2.5 offers more characteristics that are typical for organizations with a focus on one of those perspectives. These characteristics are a kind of "dimensions" that are helpful to assess a company (or a department of a company).

In Table 2.5, the cumulative character is applied again (to positive, but not to restrictive descriptions). As an example: *product or service quality* (*Π1*) will always be a natural aspect or consequence if *production quality* (*Π2*) is the focus of the quality management. Both are essential elements of *organization quality* (*Π3*), while all three are necessary in order to provide *societal value* (*Π4*).

2.4 The FFEA System

2.4.1 The Five Modules of Future-Focused Entrepreneurship Assessment

On the one hand, the FFEA system is based on the four perspectives described in Section 2.3. On the other hand, it has its fundaments in a view on the main roles of a company within society and the natural environment. (Please remember: wherever in this chapter the word "companies" is used, all kinds of organizations are implied, unless otherwise stated, including commercial companies and non-profit organizations such as schools, healthcare institutions, government departments, NGOs, etc.)

These four main roles are called "pillars," and they are shown as pillars in Figure 2.5.

The left pillar, *operations*, deals with all internal processes that enable the company to operate, e.g., the human resource management (HRM), the financial administration, the maintenance of buildings and equipment, and the procurement.

The second pillar, *primary process*, is related to the main processes with which the company earns its income, i.e., industrial processes, healthcare (for a hospital, etc.), education (for a school or university), etc.

The third pillar, *innovation*, has to do with all kinds of change processes, ranging from modest adaptations to thorough innovations or transformations, including fundamental or applied research. They may be related to the primary processes, but just as well to marketing policies, operations, financial management, or the company identity itself. Redefining the company mission takes place through an exchange of this pillar with the identity.

Table 2.5 More characteristics of the four perspectives

Perspective	II1: Diligent	II2: Targeted	II3: Systemic	II4: Holistic
Motto	**Traditional:** "We do it as we always did, it has always worked well."	**Structured:** "We work according to well-considered structures."	**Flexible:** "The structures are there for the employees, and not the other way around. Where necessary, we customize our activities."	**Adaptive:** "We work as the circumstances require, we always adapt our structures if necessary: we are a living organism."
Definition of "quality"	Product (or service) quality	Production quality	Organization quality	Added societal value (stakeholder value)
Mission, strategy, policy	**Operational:** The management is primarily focused on the short term, in order to be able to carry out the ongoing processes. Mission and strategy do not play a clear role, if they even exist.	**Tactical:** The policy is aimed at achieving medium-term concrete goals, taking into account foreseeable internal and external developments in the next few years.	**Strategic:** The mission is regularly reviewed and adjusted as necessary or even redesigned. The strategy is derived from the mission, and focuses on the long term. The policy is derived from the strategy.	**Visionary:** The prevailing ideas about the organization, its mission, and its products or services are constantly challenged actively and creatively, with inspiring views not being shunned.
Decision making and involvement	**Hierarchical:** Many decisions are taken top-down, without much	**Supported:** Decisions are taken by the management after the level of	**Participative:** Employees ranking from high to low in the organization are	**Open:** The outside world is also actively involved in processes that lead to decision

(Continued)

Table 2.5   Continued

Perspective	II1: Diligent	II2: Targeted	II3: Systemic	II4: Holistic
	consideration for support or resistance of employees or lower management. Although it may seem different, the management actually holds every responsibility in its own right. Bottom-up processes are not appreciated.	support of the staff has been determined. In case of insufficient support, decisions are adjusted or postponed, or staff involvement is increased.	genuinely involved in processes that lead to decision making. They experience shared responsibility, a *sense of ownership*.	making. These include, e.g., customers, suppliers, and civil society organizations. The organization is a strong brand, and is experienced by the outside world to some extent as "of all of us."
Initiatives	**Ad hoc:** Initiatives usually arise ad hoc, either by individual employees without management support or by management without the support of the employees.	**Supportive:** Employee or departmental initiatives are encouraged and supported by the management, if they contribute to realizing the goals of the organization.	**Sensitive:** Possible trends and developments in the coming decades are being actively studied, and conclusions from this influence the company strategy.	**Interactive:** Initiatives are established through intensive interaction with the outside world, e.g., with clients and civil society organizations, and have effects on both the organization and the outside world.
Innovations	**Glued-on:** Innovations are glued-on without questioning the existing processes, thanks to which not much really changes.	**Accepting:** Innovations that are helpful in achieving the goals of the organization are embraced and applied by management and staff, and incorporated into existing structures and processes.	**Embedded:** Innovations are embraced in such a way that they become a part of the nature and the identity of the organization.	**Trendsetting:** The organization is trend-setting with its innovations, thereby conscientiously contributing to societal developments.

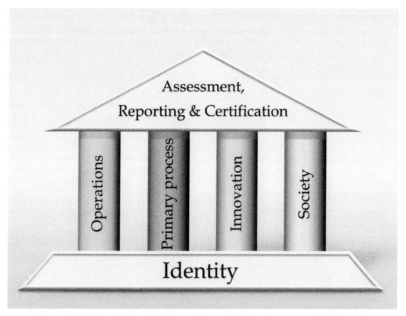

Figure 2.5 The "Temple": four pillars, their fundament, and their roof.

The fourth pillar, *society*, contains all kinds of contacts and relations of the company with the outside world, including the local, regional, or even global community and the natural environment. Low-ambition relations may involve donations by the company to charities or small-scale actions of members of staff to local schools or playgrounds. More ambitious interactions within society may, e.g., range from participations in societal discussions to public–private partnerships.

In a well-organized company, the four pillars will all have their fundament in the company identity, as Figure 2.5 shows, which may be made explicit in a company mission. Besides, many companies evaluate their processes and results regularly and report about them. As a consequence, companies may be awarded some kind of certification, e.g., an ISO Certificate. Together, assessments, reports, and possibly certificates form the "roof" of the "Temple" of Figure 2.5. Actually, this figure implicitly contains a PDCA Cycle (Plan–Do–Check–Act) or Deming Cycle [23], in which the identity stands for "Plan"; the four pillars for "Do"; the roof for "Check"; and the feedback from the evaluations toward the identity – with as its most ambitious option "redefining the company mission" – for "Act."

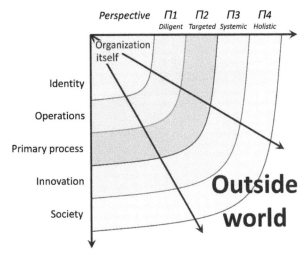

Figure 2.6 The (imperfect) correspondence between the four pillars and the four perspectives.

There is a kind of correlation between the four pillars and the four perspectives, although it is not an exact 1-to-1 match. Both correspond to a certain extent to an inside–outside scale, ranging from the very core of the organization itself to the entire outside world, as Figure 2.6 illustrates. Diligent companies (*Π1*) tend to have a focus on operational aspects (but also on the primary process, which shows this is not a 1-to-1 relation). Targeted companies do the same, but with an emphasis on measurable outcomes and processes. Systemic companies, acting from a long-term perspective, usually have a clear policy on innovations, while holistic companies are strongly embedded in society.

The four-pillar model is reflected in the structure of the FFEA system, as Figure 2.7 shows. FFEA consists of five modules: one for each of the pillars and one for the fundament, i.e., the identity.

The details of each of the five FFEA modules are explained in the next section.

2.4.2 Six Topics to Each Module

For each of the five FFEA modules, a range of six topics is distinguished, as Table 2.6 shows. Three of them are arranged along the well-known Triple P ([7], p. 31): people, planet, and profit. They all belong to the "Do" section of the already mentioned PDCA Cycle, together with a topic called "Basics,"

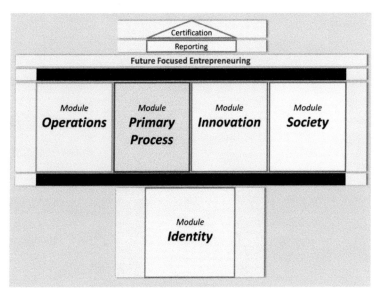

Figure 2.7 The five FFEA modules, derived from the "Temple" model.

Table 2.6 The six topics of a module

PDCA	Topic		PDCA
Check	Key values	→	Act
Do	World *(Planet)*		
	Humans *(People)*		
	Finance *(Profit)*		
	Basics		
Plan	Roots	←	

which is defined as the starting point or preparation to "Do," i.e., to operate. All of these four topics are based on a "Plan" fundament, defining the "Roots" of an organization. Finally, the "Check" part of the PDCA Cycle is expressed as the assessment of the rate to which the roots are realized, based on key values as indicators.

When all six topics are applied to all five modules, the full extent of FFEA becomes visible, as is shown in Figure 2.8. This figure clearly shows how all kinds of topics that are relevant for CSR are integrated elements of FFEA.

Thus, all in all the FFEA system consists of 6 × 5 = 30 topics. These topics, or rather a selection of them, form the agenda for FFEA assessments. Each of the 30 topics is expressed in detail, making use of the four perspectives, as will be described in the next section.

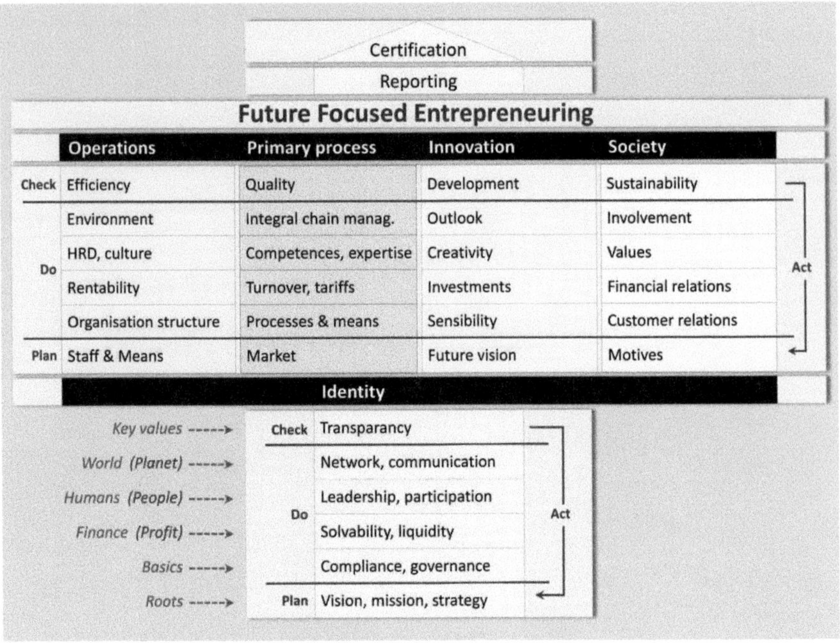
Certification
Reporting
Future Focused Entrepreneuring
Operations
Primary process
Innovation
Society
Check
Efficiency
Quality
Development
Sustainability
Environment
Integral chain manag.
Outlook
Involvement
HRD, culture
Competences, expertise
Creativity
Values
Do
Act
Rentability
Turnover, tariffs
Investments
Financial relations
Organisation structure
Processes & means
Sensibility
Customer relations
Plan
Staff & Means
Market
Future vision
Motives
Identity
Key values ----->
World (Planet) ----->
Humans (People) ----->
Finance (Profit) ----->
Basics ----->
Roots ----->
Check
Transparancy
Network, communication
Leadership, participation
Do
Act
Solvability, liquidity
Compliance, governance
Plan
Vision, mission, strategy

Figure 2.8 The five modules of FFEA.

Actually, the FFEA system has the shape of a matrix of six rows and five columns. Each column is formed by a module, while each row is formed by a "Theme Group." An example of such a theme is "Finance" (the "profit" part of the Triple P), as shown in Figure 2.9.

Before going to the final details of FFEA, with descriptions of the four perspectives for each of the 30 topics and offering the assessment procedures, the system will first be applied to a particular case.

2.4.3 The Royal Dutch Shell Case

The British–Dutch multinational oil and gas company Royal Dutch Shell PLC is the largest of the "supermajors": the dominating oil and gas companies, among which also BP, ExxonMobil, Chevron, and some more. According to the Fortune 500 list [24], it is the fifth-largest company in the world, with revenues in 2016 of US $272 billion. It is evident that its CSR and sustainability policy has a major impact on global economic and environmental developments, especially related to climate change – in either a positive or a negative sense.

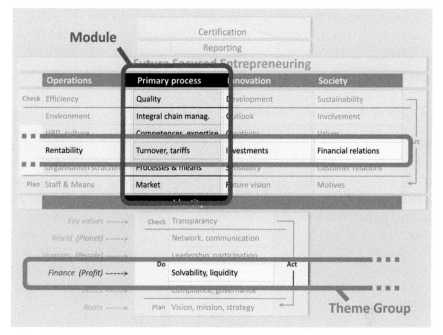

Figure 2.9 Modules and theme groups.

Already as far back as 1991, Shell was aware of the dangers of climate change. The catastrophic risks were described in a documentary produced by Shell, a movie clip [25] that for some reason was not publicly seen until it resurfaced in 2017 [26].

In 1998, Shell confirmed the dangers of global warming in its corporate sustainability report [27]:

> *"Human activities, especially the use of fossil fuels, may be influencing the climate, according to many scientists, including those who make up the Intergovernmental Panel on Climate Change (IPCC). (...)*
>
> *The burning of fossil fuels – coal, oil, and natural gas – together with other human activities, such as deforestation, releases greenhouse gases, mainly CO_2 into the air.*
>
> *Their concentration in the atmosphere has been rising since the industrial revolution. This has led to an enhanced greenhouse effect and there is concern that it will cause the world to warm*

up, which could lead to a change in climate and local weather patterns, possibly with increased droughts, floods, storms, and sea level rise. The average temperature of the earth has risen by about half a degree celsius over the last century, possibly due in part to greenhouse gas emissions caused by human activity."

In spite of this, through several decades Shell, together with other "super-majors," has attempted to deny human-caused climate change by criticizing scientific research and influencing the public opinion and political decision making. Around 1990, it became a member of the Global Climate Coalition (GCC), an international lobbyist group of businesses that opposed action to reduce greenhouse gas emissions. In 1997, the Royal Dutch left the GCC after heavy pressure from environmental groups, but it continued its lobbying, also described as deception [28].

Around 2000, Shell started a new division, Shell Solar, in an attempt to open a market for sustainable energy. To the surprise of many, the division was sold in 2006 to SolarWorld, and Shell ceased its investments in solar or wind energy.

A decade later, in 2016, a group of Shell shareholders who called themselves "Follow This" [29] prepared a resolution for the Annual General Meeting (AGM). Their "Shareholder Resolution" ([30], p. 10) started and ended with:

"As shareholders in Shell, we want you to know that we expect the company to move in a new direction. We want Shell to change course and make the move to sustainable energy. We know you can make a difference. (...)"

"We ask the board to make the right investment decisions, in the interest of all stakeholders of the company; clients, personnel, shareholders, and society. For the future of Shell."

The resolution explicitly mentions "all stakeholders," not just the *shareholders*. This corresponds directly with Table 2.1 in this chapter, in which the highest "Maslow level" mentions: Move from *shareholder value* to *stakeholder value*.

In their response ([30], p. 10), the directors of Shell unanimously recommended to vote against the resolution. At the AGM, the resolution was rejected by 97.22% of the votes [31].

One year later, the "Follow This" group, expanded with many new members, proposed a new shareholder resolution ([32], p. 6). This time, the text was modest and careful, as the group attempted to formulate a proposal that "nobody could possibly be against":

> *"Shareholders support Shell to take leadership in the energy transition to a net-zero-emission energy system. Therefore, shareholders request Shell to set and publish targets for reducing greenhouse gas (GHG) emissions that are aligned with the goal of the Paris Climate Agreement to limit global warming to well below 2°C.*
>
> *These GHG emission reduction targets need to cover Shell's operations as well as the usage of its products (...), they need to include medium-term (2030) and long-term (2050) deadlines, and they need to be company-wide, quantitative, and reviewed regularly.*
>
> *Shareholders request that annual reporting includes further information about plans and progress to achieve these targets."*

Although – instead of being imposed to accept targets for the company – the directors were invited only to define their own targets, their response ([32], pp. 7–8) was:

> *"Your Directors consider that Resolution 21 is not in the best interests of the company and its shareholders as a whole and unanimously recommend that you vote against it. (...) The resolution is unreasonable with regard to what the company can be held accountable for and would be ineffective or even counterproductive for the following reasons: (...)*
>
> *We are convinced we have all the required flexibility to adapt and remain relevant and successful, no matter how the energy transition will play out. We believe that by tying our hands in the early stages of this evolution, this resolution would weaken the company and limit our flexibility to adapt. (...)*
>
> *We are pursuing at scale the development of new gas markets around the world. Putting limits on these business activities, which do indeed increase emissions by Shell and its customers but reduce emissions in the system overall, would be counterproductive. (...)*
>
> *To impose targets on a single supplier in this complex system does not only fail to address the actual challenge (as it will not reduce system emissions overall because customers will simply turn to alternative suppliers); it would also undermine our ability to play*

*an active role in the transition and would hinder long-term value
creation for the company and its shareholders."*

This time, the resolution was rejected with 93.66% of the votes [31].

Several elements in the directors' considerations are striking.

One is that the reference to "alternative suppliers" reminds of the so-called "Prisoner's Dilemma" ([7], p. 229). It appears that Shell mentions the competition as a reason, not to adopt a leading position toward sustainable energy.

Another is the mentioning of the shareholders, twice. This seems to indicate that a transition from shareholder value to stakeholder value, which is occurring in many small and large companies ([7], p. 343), has not yet reached Shell.

This negative attitude of Shell's directors toward an innovative, leading position, an attitude that was accepted by a large majority of its shareholders, is all the more awkward as the global energy market is changing rapidly. Costs of several kinds of sustainable energy are decreasing at a fast pace, even exponentially, as Figure 2.10 shows.

Although the costs of fossil energy are presently low (but highly fluctuating), it is to be expected that sustainable energy will outcompete coal,

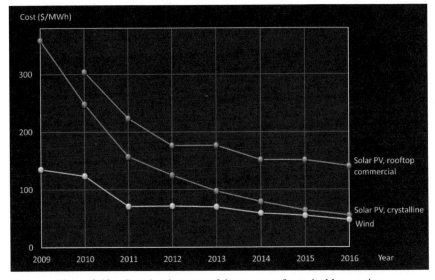

Figure 2.10   Cost development of three types of sustainable energies.

Source: [33].

oil, and gas within a limited number of years – in some circumstances, it already does. Presently, many companies are investing heavily in it, and they are prospering: a scenario is unfolding which reminds of the Eastman Kodak case in Section 2.2.

It is fascinating to compare the history and the possible future of Shell with the "Bridges Model" for organizational development. William Bridges based his model [34] on the Myers–Briggs model for personality types [35], applying it to organizations instead of individuals. A short description of the Bridges model is offered in [36]:

> *"The life of an organization usually follows a typical series of phases. In order to characterize these phases, Bridges makes use of four dichotomies with which organizations can be categorized. Only one will be described here: it is the distinction between "extraverted" organizations (focusing on the outside world, with open boundaries, quickly reacting to external changes) and "introverted" organizations (focusing on the internal processes and procedures, with closed boundaries, reacting to external change only after careful reflection). (. . .)*

Making use of this dichotomy, the following natural organizational development phases can be described.

1. *The dream. An idea or ideal is in someone's mind, nothing more than that yet. The phase is introverted.*
2. *The venture. Action is taken to realize the dream. This demands a lot of creativity and involvement of the (small) staff. In this phase, there is an intensive interaction with the outside world, and the phase is extraverted.*
3. *Getting organized. The success of the organization increases, and so does the size of the organization, which becomes more complex. This demands a set of fixed procedures and some standardization. In this phase, the organization is introverted, trying to develop its structure.*
4. *Making it. Now that the internal processes are all right, the organization is able to grow further. This demands the discovery of new market opportunities, and so the organization is extraverted again, as account managers swarm the earth.*
5. *Becoming an institution. Some organizations may exist for a long time and become a part of the "establishment." Its name is known to many. It may even be the dominant market leader, able to set the standards*

that others have to follow. In this phase, the urge to renew the organization or its products and services may shrink, since the organization "knows" what the customers want. Therefore, the tendency is to become introverted again.

6. *Closing in. This is actually a deepening of phase 5. The bureaucracy weighs heavy on the organization. The introversion gets so strong that customers are made to feel that the organization is doing them a favor by serving them. The organization is a "rusty mammoth."*

7. *Death. Because the organization has virtually lost its ability to change, the very first significant external change will kill it. Its place is taken by a dozen new extraverted ventures, based on new dreams, and so one generation disappears and makes place for the next one.*

Many variations exist to this natural development process. Organizations in phase 6, or any earlier phase, may be revitalized and start a "second life," perhaps even entering an iterative process of repeated rejuvenation. Organizations in phase 4 may meet a sudden change (a trend break) in the external circumstances, and die. But generally, the described development pattern is recognizable in many existing and past organizations."

When applied to Shell, the Bridges model raises the suspicion that the Royal Dutch is somewhere in phase 5 or more likely 6. Has Shell become a "rusty mammoth"? Bridges writes about phase 5 ([34], p. 72): "Radical new ideas are distrusted because everyone agrees that the status quo is pretty darn good." And about phase 6: "Discordant information from the outside – such as increasing numbers of complaints or the news that a competitor is launching a new product line – gets filtered out or watered down so that it is of little concern."

Applying Tables 2.3 and 2.5, together characterizing the four perspectives of companies, a lot more can be concluded about Shell. The company certainly does not operate from a holistic ($\Pi4$) perspective: the company can hardly be called "visionary"; it certainly does not "participate in creating or supporting developments (gates) toward sustainable, societally preferred futures (utopias)." Decision making and involvement cannot be seen as "open," and the company is by no means trendsetting.

Table 2.5 defines "quality" for holistic ($\Pi4$) organizations as "added societal value (*stakeholder value*)." The Shell directors' response in 2017, with its emphasis on shareholders' interests, again does not indicate that Shell is a $\Pi4$ company.

It would be interesting to formally assess the Shell Company with FFEA, using the procedures that will be described in the next section. But even without such a formal assessment, the FFEA system as shown in Figure 2.8 offers a lot of information.

In the *Innovation Module*, Shell evidently misses sufficient sensibility for trends in the outside world; a long-term perspective; and genuine creativity. Its investments are in the wrong direction: "new gas markets around the world."

In the *Society Module*, the company is low on societal involvement, societal values (only mentioning: "value creation for the company and its shareholders"), and sustainability.

In the *Identity Module*, a clear vision and a sense of leadership cannot be observed, and the communication with stakeholders is mainly in one direction: the board seems to be "audibly challenged." Even a basic compliance seems to be lacking, as the company is accused of deception [28] and of systematically violating laws [37]. (In Table 2.1, compliance is explained as not just related to laws and regulations but also to corporate codes and ethical standards.)

If the FFEA theme group *Finance* is applied, there may be dark clouds gathering above Shell. All the above-mentioned weaknesses of the "rusty mammoth," among them wrong investments, may soon lead to: decreasing turnover; negative rentability; zero liquidity; insufficient solvability; and finally, Bridges' phase 7: death.

Table 2.2 puts the oil companies in the category "vulnerable." Maybe this description is too modest for Shell. Perhaps "threatened" or even "endangered" would be more appropriate. Instead of redefining the company mission, Shell remains to see itself as a "fossil company." A self-fulfilling definition?

Royal Dutch Shell received a draft of the present chapter and was asked for a reaction. Shell's response was [38]:

> *"Shell's business strategy is resilient to the envisaged implementation of the Paris Agreement on climate change, which is now progressing through the Nationally Determined Contributions. For example, the increased use of natural gas in place of coal has contributed to the emissions reductions being seen in the USA and the UK and the emissions plateau in China. Natural gas now makes up over half of the Shell portfolio. At this stage, however, industry is still facing significant uncertainties as to how government policy and consumer behavior will ultimately shape the evolution of the energy system and which technologies and business models will*

prevail. We believe we are unique in having a broader set of business options under technical and commercial development than any other company in our sector. We are pursuing these options with conviction and commercial realism. We are convinced we have all the required flexibility to adapt and remain relevant and successful, no matter how the energy transition will play out."

Whether it is true, as Shell's spokesman e-mailed, that Shell is resilient to the implementation of "Paris" will become clear in the future. At present, one conclusion can be drawn from Shell's answer: the company deals with climate change in a reactive, not in a proactive way; and – focusing on natural gas – in a traditional, not in an innovative way. Not a word is spent on societal or environmental considerations; it only seems relevant that Shell claims to remain "successful" in the future. The idea that the success of a company depends on the success of society as a whole appears to be missing.

This confirms the conclusions that Shell is not by far a $\Pi 4$ company. From one of the largest companies in the world, leading in causing climate change, a different role was to be expected: leading in combating climate change as well.

The Royal Dutch Shell may not be ready for the future.

All in all, the Shell case proves how CSR – or rather the lack of it – leads both to a severe threat to the company, and to society and the natural environment: the negative variety of formula 1 with which Section 2.2 ends. In an adapted shape, this formula becomes:

Formula 2:
Lack of CSR = Threat to the future of the company
= Threat to the future of society and the planet

2.4.4 Details of the FFEA System

The Shell case illustrates how the FFEA system may be used to investigate the future resilience of a company. The way in which this was done did not yet make use of the full potential of FFEA, as the four perspectives were not explicitly used. Now, they will be added to the system.

For each of the 30 topics, the four perspectives have been made explicit. This chapter will not offer all of these 30×4 descriptions, but limit itself to just one example.

This example belongs to the Innovation Module, for which the initial "*N*" is used, from "inNovation." (The "*I*" is used for the *Identity Module*.) The six *N* topics are taken from Figure 2.8 and are shown in Table 2.7:

Table 2.7 The six topics of the inNovation Module

PDCA	Topic		PDCA
Check	*N6*: Development	→	Act
Do	*N5* (*planet*): Outlook		
	N4 (*people*): Creativity		
	N3 (*profit*): Investments		
	N2: Sensibility		
Plan	*N1*: Future vision	←	

As an example, the descriptions are shown of the four perspectives for *N6*: Development. In Table 2.8, first a general description of *N6* is presented,

Table 2.8 Example: the four perspectives related to *N6*: Development

General description	*Sensibility, investments, creativity,* and *outlook* – the criteria *N2* to *N5* – together determine the innovative strength of an organization. However, a great innovative force does not automatically have a big effect from that power. An elephant experiences a large gravitational force, but if it stands on a solid horizontal plane, this force does not mean that the elephant is set into motion. *Development* is the term used to indicate the actual effect of the innovative power. This topic is dedicated to what is really changing due to that power: inside and outside the organization.			
Perspective	**Π1: Diligent**	**Π2: Targeted**	**Π3: Systemic**	**Π4: Holistic**
Policy	Short term: *Operational*	Intermediate term: *Tactical*	Long term: *Strategic*	Panoramic: *Visionary*
Description	• The organization is careful and emphasizes the present good, based on the expectation that small development steps are sufficient to maintain the status quo in the future. • The company's reputation is one of solidity,	• The company has a clear business sense. Structural investment is being dedicated to new developments, which lead to a regular expansion of the product or services portfolio. • Planned In advance, staff are deployed for these innovations, or external	• The organization is constantly evolving, dealing with insecurities with agility and success. In addition, the creativity of the employees is deployed freely, supported by substantial investments. The organization does not shun a pioneer role.	• The organization succeeds in actively participating in a process of transition, which has profound consequences for the organization itself, its business sector, and society as a whole. • This transition process demonstrably

(Continued)

Table 2.8   Continued

providing products or services of consistent high quality to which the customer base relies.	expertise centers are used. • Partly due to this, the organization has been able to show growth for a <u>number of years</u>.	• The change processes have a profound influence on the products or services provided, processes and methods, staffing and organizational structure and culture, and even on the identity and mission of the organization; or at least on many of these aspects. • The change processes partly have the character of an exploration or adventure, and repeatedly lead to profound <u>transformations</u> of the organization.	contributes to significant improvements in the human and natural environment.

Notes

Number of years: Think of a minimum of 5 years in which growth in most years, and on average over that period, was demonstrable with regard to, for example, revenue, market share, and/or quantity of products or services delivered.

Transformation: A thorough, lasting change process in which goals and structures change shape and content.

Transition: A comprehensive shift in society, which changes not only the actual structures but also the way in which the world is experienced, coupled with <u>paradigm</u> <u>shifts</u> and system innovations. Historic examples include: the Renaissance, the Industrial Revolution, and the Introduction of ICT society.

Paradigm: Fundamental word, image, or concept that makes the world understandable and subjectively controlled.

Paradigm shift: Development in which certain paradigms, which previously determined the prevailing worldview, disappear or change significantly, to make way for whole new paradigms.

and next the four perspectives. Some of the concepts used in the descriptions may not be immediately clear to all users, or may be explained in different ways in various contexts. So, for these <u>underlined items</u>, explanations about their use within FFEA are offered in the notes in the bottom row of Table 2.8.

2.5 Application of FFEA

2.5.1 Assessment Principles

The FFEA assessment can be applied to every kind of company. Again: wherever in this chapter the word "company" is used, all kinds of organizations are implied, unless otherwise stated, including commercial companies and non-profit organizations such as schools, healthcare institutions, government departments, NGOs, etc.

The FFEA assessment does not necessarily have to be applied to the entire company at once. It can just as well be used to assess a part of a company, e.g., a department, a faculty, a division, a plant, a team, etc.

On the other hand, FFEA can be applied simultaneously to a group of companies that have some kind of cooperation, either long-term and structurally or just for a limited period or intensity.

For all of those, the neutral term "organization" is used.

The selection of topics

In most assessments, not all 30 topics are discussed. Often, a combination of two or three modules is chosen, e.g., *Identity*, *Primary process*, and *Innovation*. Alternatively, one or more theme groups (see Figure 2.9) may be selected, such as the combination: *Roots*, *World*, and *Key values*. An assessment of three modules (18 topics) or three theme groups (15 topics) will probably take one day.

Also, a special selection of topics out of several modules and theme groups may be assessed, based on specific needs and desires of the organization. One of the cases described in the next section will illustrate this.

The application of FFEA should preferably not be an isolated, one-time incident within an organization. The value increases hugely, if it is adopted as an integrated part of a system of quality management. In order to make this possible, it is recommended that the *Identity Module* or at least those topics of it that are considered as essential for the organization in its present state are included in the assessment. This guarantees that the mission of the organization is an integrated part of the process, allowing to think freely and creatively about it and enabling all kinds of strategic or operational issues to the mission.

The Consensus Group

Conclusions about topics that FFEA deals with can never be drawn in a 100% objective way. By their very nature, they are subjective concepts that cannot be measured in an exact way. That is the reason why the four perspectives are expressed on an ordinal scale, and not, e.g., on a ratio scale. Nevertheless, if an assessment would do no better than render subjective results, it would be mostly useless.

Future-Focused Entrepreneurship Assessment operates between those two opposites: its outcomes are *intersubjective*, resulting from a decision process based on consensus building.

For this purpose, a group of persons is composed, whose members together are representative for the organization itself and all of its relevant stakeholders.

At the very least, the following need to be members of the Consensus Group:

- Managers/directors/board members, etc., i.e., one or several persons with authority or mandate for the organization;
- Members of the personnel;
- Customers/clients/patients/students, etc., i.e., those who benefit from the primary output (products or services) of the organization.

Other members of the Consensus Group may be, e.g., members of an advisory or a supervisory board; representatives of financial institutions related to the organization; representatives of local, regional, or national governments; geographical neighbors; suppliers; members of societal or environmental interest groups; and so on.

Even external philosophers, experts, visionaries, or artists may be invited.

The relevance of these representatives varies, e.g., depending on the kind of organization; the level within a larger company; the kinds of answers and insights it is looking for; and the selected topics. If several modules or theme groups are applied within one assessment, the composition of the consensus group may vary for each module or theme group.

In total, the group should count somewhere between 8 and 15 members.

From the present to the future: FFEA as a strategic instrument

During the assessment, for each of the selected topics, two decisions will be made by the Consensus Group.

The first decision is about the *present state*: from which of the four perspectives does the organization act, considering the topic that is being discussed?

The second decision deals with the so-called *desired state*. Considering the current topic, which perspective should be dominant in the nearby future?

In order to make the discussions about the desired state as concrete as possible, an exact year and date is appointed for this desired state, as a preparation to the assessment. The date may be, e.g., 1 or 2 years after the assessment; or at the end date of the current strategy plan; or when the next external accreditation audit takes place; etc.

2.5.2 The FFEA Assessment

The actual assessment consists of three parts.

In the first part, the assessor (who may be an external, certified FFEA assessor or a member of the organization staff, e.g., a quality manager with a training in the application of FFEA) explains to the Consensus Group the structure and purposes of FFEA, and the assessment procedure that just started.

2.5.2.1 Individual scoring

In the second part, the members of the Consensus Group receive the detailed descriptions of the four perspectives of the selected topics (like the ones in Table 2.8). They are invited to form their own personal opinion about the perspective ($\Pi 1$–$\Pi 4$) the organization is in at present, with respect to each separate topic. During this part, they are not allowed to communicate with the other participants, or to consult any documents; at this stage, it is their personal, subjective opinion that matters.

The participants are allowed not only to choose between $\Pi 1$ and $\Pi 4$, but also for what is indicated as $\Pi 0/?$ This is the score they choose, if they either are convinced that even the demands of $\Pi 1$ are not met, or if they *just don't know* if they are met.

After all participants decided about their scores, the results are gathered. The results are anonymous, apart from the fact that categories (such as Managers, Customers, etc.) are shown in the overview. For module P (primary process), such an overview (a "cloud of opinions") may look like Table 2.9.

2.5.2.2 Consensus meeting

The third part of the assessment, the consensus meeting, is the main part. Whereas the first two parts together may take about an hour, the consensus meeting may take the rest of the day (if, e.g., three modules are selected).

At the start of the meeting, the participants receive the "cloud of opinions," so they are aware of the variance in viewpoints.

Table 2.9   A possible result of individual scoring of Module P: Primary Process

Perspective	Π0/?	Π1: Diligent	Π2: Targeted	Π3: Systemic	Π4: Holistic
P6. Quality	PCO	PPPMCCA	PPPM		
P5. Integral chain management	C	PPPMCCAO	PPPM		
P4. Competences and expertise	P	PPPPACO	PMC	M	
P3. Turnover and tariffs		PC	CAO	PPPMMC	PP
P2. Processes and means	CCC	PMAO	PPPPPM		
P1. Market	O	PPPCCC	PPPMMA		

P, Personnel; M, Manager; C, Customer; A, Advisor; O, Other.

One by one, the selected topics are discussed, with the FFEA assessor as the chair person. For each topic, the group decides first about the present state and next about the desired state. Conclusions are *never* based on voting, but always on consensus; it is up to the assessor to see to it that a genuine consensus is reached and explicitly concluded.

The process of the assessment looks like outlined in Figures 2.11–2.14. As an example, Figure 2.12 shows that, with respect to topic *P1* (Market), the organization in its present state acts from perspective *Π2*, and should reach *Π3* at the appointed date of the desired state. Figure 2.13 shows that topic *I4* (Leadership and Participation), for which the organization is in *Π3*, the consensus group agrees that no improvement is desired.

These conclusions are not just formulated in the form of numbers, but also – even primarily – in words, expressing why and how. It is the task of the assessor to guarantee that the decisions are based on concrete arguments, and that it is made explicit how the desired state is to be realized. Thanks to this, a consistent *reality check* is an integrated part of the assessment: based upon it, the participants are confident that the desired state can actually be realized within the appointed time frame.

Notes are taken during the assessment about the argumentations and conclusion. So, at the moment the assessment finishes, the report is complete as well, apart perhaps from a bit of editing.

After all selected topics are completed, the consensus group appoints a small number of highest priorities, i.e., of those improvements that are considered as crucial to the organization development within the appointed time frame. Selecting the priorities usually takes just a few minutes, as all topics have been discussed thoroughly already. After this, the assessment is finished.

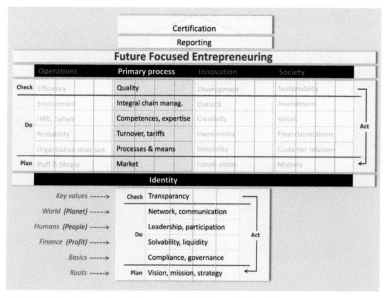

Figure 2.11 The selected modules are highlighted; the others are grayed-out. In this example, Identity and Primary process are selected.

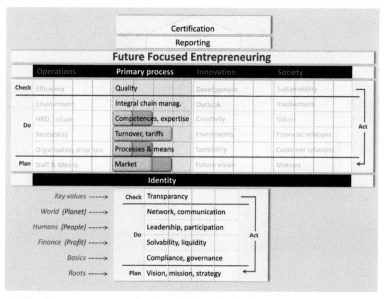

Figure 2.12 The consensus meeting is at full speed. The present state and the desired state have been decided in consensus for the first four topics of the primary process module.

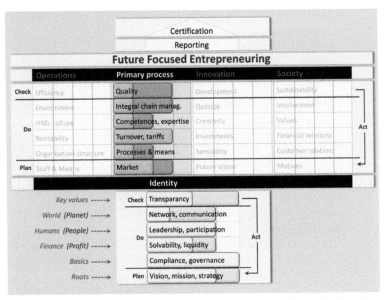

Figure 2.13 The consensus meeting is completed. After the high priorities have been selected, the assessment is finished.

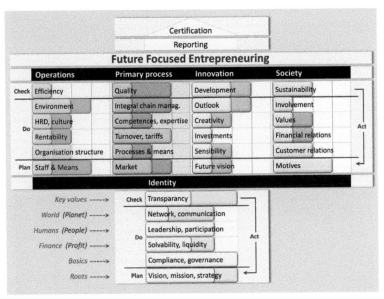

Figure 2.14 If all modules are assessed – perhaps in the course of a series of assessment of the same organization within a few weeks or months – the overall result may look like this.

2.5.3 The Results, or: What You Get

A FFEA assessment is an investment. The main costs are probably due to the time that the members of the Consensus Group spend; but these may not be out-of-pocket costs. On the other hand, an experienced FFEA assessor is essential. An external assessor will have to be paid; an internal one will have to be trained by external experts first, which also involves expenses.

So, the question is fair: what does it get you? The answer, in six parts, is shown in Table 2.10.

As to the strategic plan, mentioned in Table 2.10 as WYG 4: it is important to emphasize the informal status of the consensus meeting. During that meeting, a lot of concrete conclusions have been drawn. However, these conclusions are no formal decisions, and they should not be, too. A variety of stakeholders together have come to these conclusions. Some of the participants are external persons; others are members of the personnel who don't carry any decisive responsibility for the organization.

It is up to the managers, directors, or board members who carry formal responsibility or have a mandate, to make final decisions. However, it is impossible to make such decisions during the assessment itself. Hence, in order to benefit from the assessment, a management meeting is to be arranged soon (e.g., on the next day) after the assessment. At this meeting, the FFEA assessor usually is present: this time not as the chair, but as an expert advisor. During the meeting, the formal decisions are made, based on the outcomes of the assessment. This will probably not be a very difficult task, as the

Table 2.10   What You Get (WYG) out of a FFEA assessment

What You Get	Explanation
WYG 1. Report	A well-structured report of the conclusions and argumentations that led to them, including a nice picture like Figure 2.14.
WYG 2. Insight	Insight in the present state: intersubjective conclusions, based on the consensus of a well-selected group of participants.
WYG 3. Plans	Concrete plans to realize a desired state, creating a future that the consensus group believes in.
WYG 4. Strategy	Input for the creation of a strategic plan for the next years.
WYG 5. Involvement	Involvement, a feeling of "being heard," support, enthusiasm, and a shared sense of ownership of all those who participated. Next, also of many others, after they hear the stories of the assessment through the mouths of their colleagues.
WYG 6. Resilience	The assessment will certainly contribute to plans and energy that both will increase the future-proof resilience of the organization.

responsible persons (or at least a delegation of them) have participated in the assessment, during which conclusions were drawn based on consensus by all.

So, the main task of the management meeting is to elaborate on the assessment outcomes, translating them into a genuine strategic plan, including such things as: a time schedule; persons who will be given responsibilities for parts of the plan; facilities for those persons, including, e.g., time, budget, tools, mandate, training, etc.; and a system of evaluation and reporting.

2.5.4 MSPOE: From Mission to Strategy to Policy to Operations to Evaluation to Mission

After the strategic plan has been designed and decided, it is to be translated into one or more policies, and next to concrete operations that are performed in the time period between the "present state" and the "desired state." This process is "zooming in": a shift from a long-term to a short-term approach, just like the *RESFIA+D* competence *F1* (Section 2.3) describes: *Think on different time scales: flexibly zoom in and out on short- and long-term approaches.* If the strategy plan is based on the mission of the organization – if the *Identity Module* was applied – this zooming in is even more evident: see Figure 2.15.

One or more years later, when the date of the desired state nears, it is the right time to plan an evaluation of the strategic plan. This should be done through a new FFEA assessment. Actually, this is a nice example of the already quoted PDCA Cycle, where the FFEA assessment is the

Figure 2.15   Zooming in from Mission to Strategy to Policy to Operations.

"Check," followed by the management decisions ("Act") that lead to a next strategic plan.

The transition from the operations to the evaluation and next to the mission is "zooming out," which not only completes the quality cycle but also the *RESFIA+D* competence *F1*.

All in all, this zooming in and out, if repeated several times, can be compared to a breathing motion: *breathing in and breathing out, breathing in and breathing out*. For companies, FFEA refers to this breathing motion as *MSPOE*: Zooming in (*from Mission to Strategy to Policy to Operations*) and zooming out (*to Evaluation, and then back to the Mission again*).

"MSPOE" is pronounced as a word that more or less rhymes with "disco," but with a "silent i."

The MSPOE principle is illustrated by Figure 2.16, where every occurrence of "Evaluation" stands for a FFEA assessment.

Figure 2.16   Zooming in and out with MSPOE.

2.6 FFEA Case Studies

The application of FFEA will be illustrated with two case studies, differing considerably.

2.6.1 The Tilburg Mentaal Case

Tilburg Mentaal is an institution for mental healthcare in the Netherlands. A number of psychiatrists and psychologists work for the organization, as well as assistants, technicians, a secretariat, and others. The management is formed by the top psychiatrist and psychologist who are the owners of the institution.

In the Netherlands, there is a lot of uncertainty for institutions like these. This is caused partly by the need, expressed by the government, to reduce healthcare costs in the country considerably. This is effectuated mainly by putting pressure on the health insurance companies, who pay for nearly all of the budget of the healthcare institutions. Mental health-care especially feels the consequences of this "cheese-slicer" approach, as it is called. This caused a lot of uncertainty for Tilburg Mentaal, a few years ago.

Other factors also had a lot of influence. The growing self-awareness of the patients of the institution changed the customer relation with the institution. The heart-felt wish to operate in a more sustainable and socially responsible way also contributed to a feeling that it had become urgent to assess a number of topics as a preparation for a reorientation.

A FFEA assessment was prepared by the managers and an external assessor. Together, they came to the conclusion that the highest priority was to be put on the *P* module (Primary process) and the *S* module (Society). It was estimated that, if the relation between both aspects ("*What they want*" and "*What we do*") was redefined and optimized, the financial situation could be improved as a next step.

The Consensus Group was formed, the assessment took place. The results of the individual scoring are shown in Table 2.11. The "Others" were: a representative of a healthcare insurance company and the accountant of the institution.

During the consensus meeting, conclusions were drawn. They are shown in Figure 2.17.

Table 2.11 Tilburg Mentaal: Individual scores for the present state

Perspective	Π0/?	Π1: Diligent	Π2: Targeted	Π3: Systemic	Π4: Holistic
		Module S: Society			
S6. Sustainability		SMMPVO	SPP	SO	
S5. Involvement	PO	SMV	SP	SMO	P
S4. Values		P	SPPO	SSMMVO	
S3. Financial relations	SPPP		SMVO	SM	O
S2. Customer relations		SPPV	MM	SSOO	P
S1. Motives		PPO	SMV	SSMPO	
		Module P: Primary process			
P6. Quality			MV	SSSPOO	MPP
P5. Integral chain management	PPOO	SM	SP	SMV	
P4. Competences and expertise		P	POO	SSSMPV	M
P3. Turnover and tariffs	PO	SSMPO	SV	MP	
P2. Processes and means		SP	SPPV	SMO	MO
P1. Market		P	SSPPVO	SMMO	

S, Staff member; M, Manager; P, Patient; V, Supervisor; O, Other.

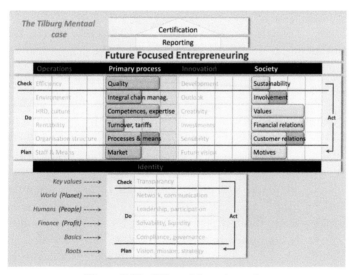

Figure 2.17   Tilburg Mentaal: results.

Here are a few quotes from the report:

P1, Market:

Present state: "It is unclear who exactly are the customers. Are they the patients we help? Or the health insurance company that pays us? Or the referrers, such as the family doctors?

In our current way of working, we are not capable of long-term thinking, as the national laws and regulations change all the time. This is also caused by the every-day working pressure."

P2, Processes and means:

Present state: "Tilburg Mentaal works methodically and well- structured. The staff participates in formulating the protocols, we are really self-organized. *Just in time* is difficult, due to the waiting lists."

Desired state: "The communication with patients about the waiting lists can be improved, partly by creating a "safety net" for the period between the intake and the start of the treatment. This can be done tailor-made: for each individual patient, we have ways to help them in-between, eventually with the help of other institutions. Individual feedback to the referrers will also help."

S5, Involvement:

Present state: "There is no regular contact with interest groups; it is not our focus. Staff members who are interested in this have not yet had a chance to do something."

Desired state: "We will map the options for becoming societally involved. This may appear to be worthwhile or even important, we are open to it."

In a comment, one of the managers of Tilburg Mentaal afterward wrote:

"The FFEA methodology has given us a good insight into the strong and the vulnerable aspects of our company. Concrete improvement points have now been formulated and we will benefit from that. The method fits well with our philosophy of a self-learning organization focused on communicative self-control."

As a follow-up to the assessment, the discussion about the future of the institution openly continued in the next months. One year after the assessment, Tilburg Mentaal merged with another mental healthcare institution, aiming at combining their strength toward the health insurance companies, thus raising the future-proof resilience.

2.6.2 The Inventive Case

A few years ago, a number of small businesses had been working together incidentally and informally. Some of them were copywriters; others were, e.g., product designers, art designers, and marketing experts. All of them had one thing in common: they thought that CSR and sustainable entrepreneurship were vital.

One of them, called "The Inventive," took the initiative to investigate the possibility of cooperating in a more structured way. This might increase their combined creative power and flexibility, increase their market, and embed their desire to operate in a societally responsible and sustainable way.

When these wishes and ideas were compared to the "Maslow for CSR" model, i.e., the pyramid described in Section 2.1, it became immediately clear that the ideas of The Inventive, which were soon enthusiastically shared by the other companies, could be placed on each and every level of the pyramid, together creating a very convincing Motivation Mix.

In terms of the Bridges model for organizational development: the group of entrepreneurs was in phase 1, *"The Dream,"* and wanted to get into phase 2, *"The Venture."*

Instead of the classic "redefining," it was a case of "defining the company mission."

But how should this be done? There was not much enthusiasm for a complete merger of the companies; they all attached a lot of value to their independence. In a meeting with a FFEA assessor, it was decided that a FFEA assessment was to be performed to deliver clarity.

During the preparation, a particular, tailor-made route through FFEA was designed. Only seven topics were selected, distributed over all five modules. Three of them were in the "Roots" group, which is no wonder, as it was mainly the roots of the cooperation that had to be studied. The seven selected topics are shown in Table 2.12. As usual, the chronological order is to be read from bottom to top: the assessment started with *N1*, future vision, and worked its way to the last topic: *O2*, Organization structure.

Due to the specific situation, in which a possible structural cooperation was explored but had not yet started, there was not *present state* to be assessed, just a *desired state* to be discussed. Therefore, Table 2.12 does not show the individual scores for the *present state*, as usual, but for the *desired state*.

During the assessment, in which all nine entrepreneurs participated – one of them The Inventive – it appeared that for nearly all topics, the majority

Table 2.12 The Inventive: Individual scores for the desired state

#	Perspective	Π0/?	Π1: Diligent	Π2: Targeted	Π3: Systemic	Π4: Holistic
7	*O2.* Organization structure			EEE	EEE	EEE
6	*I4.* Leadership and participation			E	EEEEEEE	E
5	*I1.* Vision, mission, and strategy			E	EEEEEE	EE
4	*S2.* Customer relations			EE	EEEEEE	E
3	*S1.* Motives			EEEE	EEEEE	
2	*P1.* Market			EEE	EEEE	EE
1	*N1.* Future vision			E	EEEEEEE	E

E, Entrepreneur.

wished to create a cooperation at a *Π3* level: *Systemic*. But during the highly inspired Consensus Meeting, the minds shifted: as Figure 2.18 shows, many topics ended at *Π4: Holistic*. A main reason for this was the explicit emphasis on CSR and sustainability.

Some quotes from the report are:

N1, Future vision:

"We should aim at daring goals, it will help us to formulate ambitious goals."

"Each of us brings in his or her special expertise and experience."

"Flexible composition of cooperating teams."

S1, Motives:

"How explicit do you want to make sustainability? You expect a healthy and societal attitude of your customers: we won't try to attract oil companies or the cigarette industry; they will probably not come to us either."

"Develop new services, be innovative – e.g., "translate" services from one business sector to another."

"Be selective in the work we accept. Really do what you believe in. Due to that: become attractive to niches."

S2, Customer relations:

"Perspective *Π4*, because we want to involve our clients in such a way that they have the feeling they are a part of us."

"Fading boundaries between customers and us."

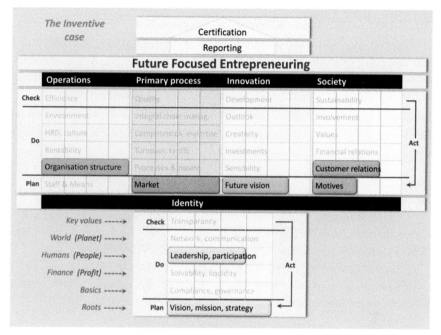

Figure 2.18 The Inventive: Results.

I1, Vision, mission, and strategy:

"A balance between idealism and realism."

"Our cooperation is to be a learning organization."

Comments after the assessment were:

"A brainstorming session on FFEA: instructive, enlightening, and stimulates a lot of thought about our business and future."

"Future-Focused Entrepreneurship Assessment brought us a good structure in our brainstorm afternoon about a new collective. One afternoon, without him, probably would have been a lot more chaotic and given us less clear conclusions."

"With the FFEA method, today we had a guided exploration about a possible collaboration, which made us quickly concrete. It is a purposeful way of engaging people in making future plans."

In follow-up meetings of the nine entrepreneurs, the basic principles formulated during the FFEA session were made operational, after which the structured cooperation started. They had gone into phase 2, *"The Venture."*

The collective was ready for the future.

2.7 The FFEA Extensions

During the MSPOE loop (zooming in, zooming out), the "E," standing for "Evaluation," is performed through a FFEA assessment. During the assessment, it will become clear which of its topics deserve special attention, as improvements on those topics received the highest priority.

Immediately after the assessment, a strategic plan is made, as was described before. In order to make this plan operational, many tools are available.

Some of these tools already existed before FFEA was developed. Examples are: the standards of the Global Reporting Initiative (GRI, [39]) and ISO 26000 [40].

Other tools were specifically designed for the operational phase, as an added instrument to FFEA. They are called the "FFEA Extensions." Each of them is aimed at one or two specific topics of FFEA: see Figure 2.19.

An overview of the FFEA Extensions is available in [41].

These extensions have the shape of a workshop, a book, a serious game, a forum, or an assessment. Some of them are shown in Table 2.13.

Figure 2.19   FFEA as a starting point for many other management methods.

Table 2.13 Some FFEA Extensions and their online sources

Topic	Extension	Type	Online
I4. Leadership, participation	STELES: Self-Test Leadership Styles	Assessment	[42]
I6. Transparency	FFEA Certificate for Organizations	Certificate	[43]
O1. Staff and means	The Seven Competences of the Sustainable Professional	Book [20]	[44]
O4. HRD and culture	RESFIA+D for HRD	Assessment	[45]
O6. Efficiency	Fundamentals of Sustainable Development	Book [21]	[46]
P4. Competences and expertise	RESFIA+D for Professionals	Assessment	[45]
P5. Integral chain management	In the Rucksack	Workshop	[47]
P5. Integral chain management	EPU – Environmental Pollution Unit	Serious game	[48]
N2. Sensibility	PopSim – Population Simulation	Serious game	[49]
N4. Creativity	Function Analysis	Forum	[50]
N6. Development	Backcasting the Future	Forum	[51]
S1. Motives	CSR Motivation Mix	Assessment	[52]
S5. Involvement	The Societal Triangle	Workshop	[53]
S6. Sustainability	The Pledge	Forum	[54]

Another important FFEA Extension, not mentioned in Table 2.13, is an assessor training, aimed at those who want to acquire the FFEA Assessor Certificate, enabling them to perform FFEA independently. Such assessments, chaired by certified assessors, may lead to a FFEA Certificate for Organizations, which is mentioned in the table related to topic *I6*.

Three of the above extensions will be discussed in more detail.

2.7.1 An Extension for Topic S1: The CSR Motivation Mix Assessment

The present chapter started with the Maslow model, applied to CSR of companies. The result is shown in Figure 2.1.

This "Maslow for CSR" model can easily be expanded into an assessment on its own right. For this purpose, Figure 2.1 is printed on a large paper (e.g., with a width of a meter or three feet), which is put horizontally on a table. A group od relevant people (managers, employees, customers, etc.) is invited to walk around the table and study the model. Next, they take two steps:

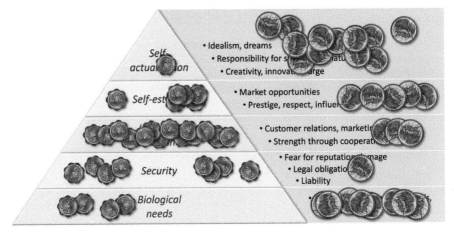

Figure 2.20   The CSR Motivation Mix Assessment.

1. Present state: Each participant receives three silver (or silver-colored plastic) coins. They are invited to place the coins on the playing board, such that the coins indicate which of the five motivation levels descibe the company motives for CSR best.

2. Desired state: Now, everyone receives three golden coins. They are requested to put them on the board in such a way that they correct or expand the present state in the way they think the company should be motivated.

The result may look like Figure 2.20.

After the board is filled, the group discusses the results and draws conclusions.

2.7.2 An Extension for Topic I4: STELES, The Self-Test of Leadership Styles

Another extension is linked to topic *I4*, Leadership. Leaders are invited to fill out a small questionnaire called STELES (Self-Test Leadership Styles), after which they receive an automated response, telling them their strengths and weaknesses in the shape of 12 different leadership styles: see Figure 2.21.

The Self-Test of Leadership Styles may be applied by formal leaders at every level of an organization and also by informal leaders, or by those who have the ambition to become a leader.

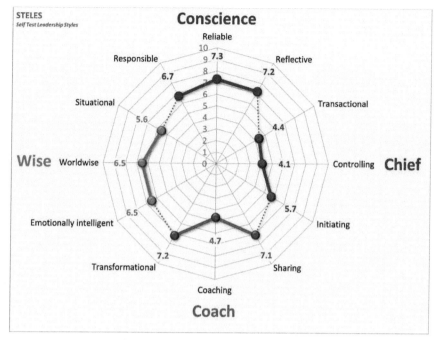

Figure 2.21 An actual result of STELES.

2.7.3 An Extension for Topics P4 and O4: RESFIA+D, or the Seven Competences

The last example of a FFEA Extension to be presented her is linked to topic *O4*, Human Resource Development (HRD) and Culture; and to *P4*, Competences and Expertise. As announced in Section 2.3, the RESFIA+D Model, also called "The Seven Competences of the Sustainable Professional," can be applied as an assessment instrument in several ways.

If it is applied by individual professionals, the result may look like Figure 2.22, which was the result of an actual result of an assessment. The figure shows how the person who did the test first filled out his personal view on his competence levels at that moment, followed by his ambition, applied to a personal development plan. Besides, some of his colleagues were invited to assess his competences, together creating a 360° feedback.

If the HRD department of a company applies RESFIA+D for all of the staff members of a team or a department, the results can be used as a policy instrument, enabling the management to design an integrated staff

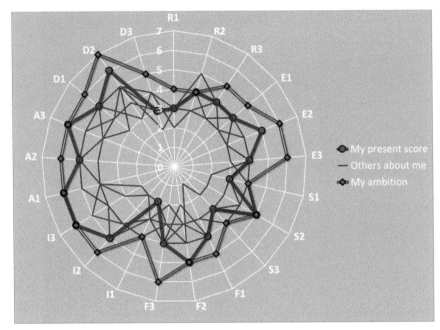

Figure 2.22   An actual result of RESFIA+D, applied by an individual professional.

development plan. The result may look like Figure 2.23. More information about RESFIA+D is published in [21] and online in [45], as mentioned in Table 2.13.

2.7.4 An Extension for Topic I6: The FFEA Certificate for Future-Proof Resilience

It does not exist yet. But preparations are being made (and may be ready when this text is read) to create a certificate that will be awarded to organizations that have proved, with the use of FFEA, to possess a high level of future-proof resilience.

A comparable certification system already exists, and has proved to be highly successful. FFEA has been developed, based on an earlier assessment system: AISHE, the *Assessment Instrument for Sustainability in Higher Education* [55]. FFEA has the same structure as AISHE, as will be described in the next section.

The AISHE Certificate is a star system: depending on the results of an AISHE assessment, a university (or a part of it, e.g., a department, a faculty,

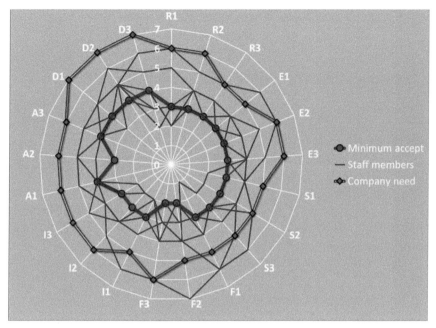

Figure 2.23 Result of RESFIA+D, applied by an HRD department.

or a campus) receives a certificate at the level of one, two, or more stars. A lot of experience has been built up with this kind of certification. Especially after the AISHE Certificate was formally recognized (see the next section), many universities have acquired one or several AISHE Certificates.

The FFEA Certificate will be defined in the same way, based on a range of one to four star levels. Certifying assessment will be chaired by certified FFEA assessors, followed by a final check by an independent certification organization.

The FFEA Certificate will be a strong indication for the likelihood that the certified organization is expected to have a sound and prosperous future. Thus, it will contribute to the trustworthiness of the organization – whether it is a commercial company, a non-profit institution, a government department or whatever – for its relations. Although a certificate of course can never be an absolute guarantee that the organization will continue to exist and flourish indefinitely, customers may rely on the certificate and be confident that the services will still be available in the next years. Banks may be confident, or at least more confident, that investments will be sound, and suppliers may expect to be paid for their deliveries.

The details of the AISHE certification, which will serve as a model for the FFEA certification, will be described below (see Section 2.8.3).

2.8 Origins and Theoretical Backgrounds of FFEA

Future-Focused Entrepreneurship Assessment was designed by the author of the current chapter. (FFEA$^{\circledR}$ is a registered trademark.) The same is true for AISHE and the various FFEA Extensions, e.g., RESFIA+D.

The FFEA system was developed on the basis of many existing theories, models, and management systems. They can roughly be divided into three groups: (1) general management models; (2) specific models for quality management, environmental management, or CSR; and (3) the AISHE assessment system, which is a model for the study and the assessment of sustainable development in higher education.

These sources will be described in this closing section.

2.8.1 Management Models

An important model that was already mentioned in the first section is of course the Motivation Pyramid of Maslow [6]. Another, which was also applied in this chapter, is the PDCA Cycle designed by Deming [23]. Both have been important to shape the description of the four perspectives, especially the detailed descriptions for each separate topic. Evidently, the PDCA model also contributed to the formulation of the six topics per module, which were described in Section 2.4.

In part, the four perspectives were also based on the Bridges model [34] for organizational development. This is the case, e.g., for topics around sensitivity, flexibility, innovation, and investments.

Also useful was the "colors model" designed by De Caluwé [56]. This model describes types of organizations and change processes taking place within them. The model offers a range of organization types, each characterized using a color. One of them is the "blueprint" organization, which is typified as focusing on clear, measurable targets (as its name suggests) and on stepwise implementation plans. Many of the characteristics of this type of organization can be found in the "Targeted" perspective ($\Pi 2$) of FFEA, which also carries notions defined by Taylor [57].

Other types defined by De Caluwé also left their traces, such as the stimulation and active participation of employees ("redprint," according to De Caluwé) and the focus on a learning organization (greenprint), which helped to define the systemic perspective (Π3), and the dynamic and complex "whiteprint," which contributed to defining the creative notions of the holistic perspective (Π4).

Sterling [58] defined a model for the levels of intensity of organizational change. These levels were helpful to define some of the characteristics of the four perspectives, especially those related to innovation processes. As an example, in Table 2.5, the row called "Innovations," ranging from "glued-on" till "trend-setting," was partly based on the Sterling levels.

2.8.2 Quality Management; Environmental Management; CSR

Several approaches to quality management have contributed to FFEA. The "classical" approach to quality management of ISO 9000 [59] was less relevant for FFEA than the quality principles of the European Foundation for Quality Management (EFQM [60]), as the latter philosophy emphasizes the importance of continuous improvement, just like FFEA and its MSPOE cycle. (Through the years, the principle of continuous improvement has also grown more relevant for ISO standards.) The EFQM model is explained, e.g., in [61].

A Dutch organization for quality management, INK ([62]), used the EFQM model as a starting point for the design of five different developmental stages of organizations [63]. To a certain extent, the four perspectives of FFEA are derived from the five stages of the INK model. The details of this will be described in the "AISHE" section, below.

Based on the INK model, several special editions of its five stages have been designed, e.g., for healthcare institutions and for educational institutions. The latter, developed specifically for the Dutch universities for applied sciences, was developed in 1999 by the so-called "Expertgroep HBO" ([64]). Details of this model, on which AISHE was based, are described in [65–67].

Several models and standards about environmental management were applied to define certain parts of FFEA: BS 7750 [68], EMAS [69], and ISO 14000 [70]. They were, e.g., used for the details of the "Environment" topic (*O5*) of the operations module and for several FFEA Extensions.

Speaking of ISO: of course, ISO 26000 [71], dedicated to CSR, was an important source for many aspects of FFEA. This is also true for the GRI standard [73] for transparent reporting, which is, e.g., related to the definition in Table 2.5 of "quality" for holistic (Π4) organizations as "added societal value (*stakeholder value*)." It is also clearly recognizable in topic *I6* of the *Identity Module*: "Transparency." If a FFEA assessment ends in the conclusion that transparency is to be one of the high priorities for realizing the desired state, GRI may offer a suitable tool.

2.8.3 AISHE: Assessment and Certification of Sustainability in Higher Education

A special source for FFEA is AISHE, which is short for: *Assessment Instrument for Sustainability in Higher Education*. Detailed information about AISHE is available online [73].

AISHE, in its first version, was developed by the author of the present chapter in 2000–2001, ordered by the *Dutch Foundation for Sustainability in Higher Education* ("Duurzaam Hoger Onderwijs," DHO). It was published in 2001 [74], after which it was applied in universities and colleges in many countries. A first evaluation of its results was published in 2004 [75].

The first version (AISHE 1.0) consisted of a list of 20 criteria, focusing on the educational aspects of an institute for higher education (HEI), and on its mission and identity, related to sustainable development. After systematic evaluations between 2001 and 2006, it had become clear that users of AISHE 1.0 wished to be able to apply AISHE also to other aspects of an HEI. The operations as well as the research were mentioned repeatedly in the evaluations.

For that reason, AISHE was thoroughly redesigned between 2006 and 2008 by an international group of universities and organizations for sustainable development in higher education, resulting in 2009 in AISHE 2.0 [55]. An analysis of the various roles of HEIs toward sustainable development gave rise to a view that is shown in Figure 2.24. The resemblance between Figures 2.5 and 2.24 is obvious.

Between 2001 and the present, the two versions of AISHE have been applied in hundreds of assessments in many countries, and as a starting point for research in universities and colleges. Examples are: Brazil [76]; Africa [77]; United States [78]; Saudi Arabia [79]; Bangladesh [80]; and Australia [81]. AISHE 2.0 is one of the assessment methods adopted by the International Platform for Sustainability Performance in Education [82]; an overview of these methods is shown in [83].

Several related assessment instruments have been derived from AISHE. One of them, called AIFSHE [84], focuses on sustainable food security. It was developed as a part of the *Comprehensive Africa Agriculture Development Programme* (CAADP) program, set up by the African Union [85].

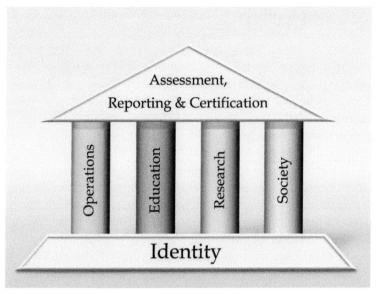

Figure 2.24   The philosophy of AISHE 2.0.

Future-Focused Entrepreneurship Assessment is another derivative of AISHE. FFEA was developed, after requests from leaders of several commercial companies reached the main developer of AISHE. The structure of AISHE 2.0 was based on Figure 2.24, resulting in five modules with six topics each. This structure was copied to FFEA, for which AISHE's educational module – which assesses the primary process of an HEI – was replaced for FFEA by a more general "primary process" module. Besides, the AISHE module "Research" was generalized to the FFEA module "Innovation."

These replacements were the starting point for the development of FFEA, which was designed, applied, evaluated, and improved between 2013 and 2017.

After ample discussions with stakeholder representatives at the beginning of the development of FFEA, it was decided to shift the focus of the newly to design assessment tool. Whereas AISHE focuses entirely on sustainable development, it was concluded that, for companies, this would be too narrow. So, the target area was widened: first to CSR, next also to other topics that have an evident relevance for the chances that a company, as a part of society in general, will be able to survive. Next to sustainable development and CSR, topics were added such as innovativity, creativity, sensitivity, societal awareness, and transparency. In the end, this resulted in the 30 topics of Figure 2.8.

Compared to AISHE, another change was made. AISHE applies the five-stage model developed by the INK, mentioned earlier in this section. During assessments made with AISHE, it had become clear that, in many ways, this appears to be one stage too many. During the assessments, participants repeatedly struggled especially with the differences between stages 4 and 5, which were diffuse to many.

For that reason, it was decided that FFEA should make use of a four-point scale, with a clear relation to different levels of policies, ranging from operational to strategic, with an extension toward panoramic. Careful comparisons between these policy levels and existing large and small companies rendered four types of perspectives of organizations, or perhaps even personalities of organizations and their relations with the future.

A lot of experience has been gained, using AISHE repeatedly, as the driving force of a repeated PDCA Cycle. Thus, AISHE has become an integrated part of the general quality management of HEIs. Based on such repeated use of AISHE, a strategy has been developed and applied within universities in order to introduce and integrate sustainable development up to a level of *"System Integration of Sustainable Development"* (SISD), as described in [86]. The MSPOE cycle of continuous improvement (the "breathing motion"), defined for FFEA, is derived from this AISHE strategic approach.

Based on AISHE assessments, in 2001, a certification system was set up, leading to the Certificate for Sustainability in Higher Education. This certificate, in the form of a star system varying from one to four stars, was formally recognized in 2007 [87] by the Dutch and Flemish Accreditation Organization (NVAO [88]). The certificate is still being awarded to universities and colleges.

Currently, preparations are being made for a comparable certificate for the extent to which companies are "Future-proof Resilient," as was told in Section 2.7.4.

2.9 Conclusion

Companies see themselves and their relations with the outside world through different perspectives. Four perspectives can be distinguished: "diligent," "targeted," "systemic," and "holistic."

Based on these perspectives, a company applies a strategy and a set of policies that range from short term (operational), through intermediate term (tactical), and long term (strategic), and finally to panoramic (visionary).

Most or all companies are not 100% consistent, varying their perspective according to the topics and issues it deals with. These topics can be divided into five groups: identity, operations, primary process, innovation, and society. These form the five modules of the FFEA management method. For each module, FFEA discerns six topics.

For every company, including commercial ones, government departments, educational and healthcare institutions, and NGOs, it is essential that its range of perspectives is inspected every few years. If not, the company is in danger of losing or lowering its *future-proof resilience*, and its health or even its continuity may be in danger.

It is the purpose of FFEA to assess this range of perspectives, enabling a company to become aware of it and to plan improvements. The improvements may vary between simple adaptations and fundamental transformations, including redefining the Company Mission.

Illustrated with case studies, this chapter concludes that the *future-proof resilience* of a company increases when it pays explicit attention to aspects of sustainable development and CSR. It also shows how this resilience of a company is closely linked to the resilience of society: depending on the size and scale of a company, this concerns a local community, or the global human society and nature, or something in between.

Future-Focused Entrepreneurship Assessment, with its development based on a range of earlier management methods, is a structured tool to investigate the *future-proof resilience* of companies. FFEA makes use of consensus meetings in which all relevant stakeholders of a company are to be represented. This guarantees that the results of an assessment are realistic and supported, and thus are suitable to be the basis for a strategic plan for improvements.

After FFEA is applied, a range of other tools and methods are available, partly pre-existing ones like ISO 26000 and GRI, and partly developed as FFEA Extensions.

References

[1] Cisco (2016). *Corporate Social Responsibility 2016 Report: Accelerating Global Problem Solving*. Available at: http://csr.cisco.com/pages/csr-reports

[2] Volkswagen (2015). *Corporate Social Responsibility and Sustainability Report 2015*. Available at: http://annualreport2015.volkswagenag.com/group-management-report/sustainable-value-enhancement.html

[3] Denso (2016). *Annual Report 2016 (Integrated Report).* Available: https://www.denso.com/global/en/csr/annual-report

[4] Unilever (2016). *Annual Report and Accounts 2016.* Available at: https://www.unilever.com/Images/unilever-annual-report-and-accounts-2016_tcm244-498744_en.pdf

[5] Royle, T. (2005). Realism or Idealism? Corporate social responsibility and the employee stakeholder in the global fast-food Industry. *Bus. Ethics* 14, 42–55.

[6] Maslow, A. (1954). *Motivation and Personality,* New York, NY: Harper.

[7] Roorda, N. (2017). *Fundamentals of Sustainable Development,* 2nd Edn. Abingdon: Routledge.

[8] MVO Nederland (2015). *Van winst naar waarde. MVO trendrapport 2015.* Utrecht: MVO Nederland.

[9] MVO Nederland (2017). *Kantelpunten Binnen Handbereik. MVO Trendrapport 2017.* Utrecht: MVO Nederland.

[10] Ryott, A. (2013). *Affärsaktivisten- Manifest för Lönsam och Hållbar Business.* Stockholm: Ekerlids förlag.

[11] IUCN (2017). *The IUCN Red List of Threatened Species International Union for Conservation of Nature.* Available at: http://www.iucnred list.org

[12] CBRE (2014). *Genesis, Fast Forward 2030. The Future of Work and the Workplace.* Los Angeles, CA: CBRE.

[13] Nowhere to Hide Beeline (2015). Available at: http://beelineblogger.blog spot.nl/2015/04/no-where-to-hide.html [accessed 2015].

[14] Eastman Kodak Company (KODK) (2017). *Yahoo! Finance.* Available at: https://finance.yahoo.com/quote/KODK?p=KODK [accessed 2017].

[15] Reina, V. (2016). *Artist, Kodak Ektachrome Panther 100 E-6. San Francisco, CA: Flickr.

[16] Siwek, S. E. (2007). *The True Cost of Sound Recording Piracy to the US Economy.* Irving, TX: Institute for Policy Innovation.

[17] Smith, M. D. and Telang, R. (2012). *Assessing the Academic Literature Regarding the Impact of Media Piracy on Sales.* Available at: https://ssrn.com/abstract=2132153

[18] Nicolaou, A. (2017). *How Streaming Saved the Music Industry.* London: *Financial Times.*

[19] Yang, L. (2016). *Music in the Air.* New York, NY: Goldman Sachs.

[20] Roorda, N. and Rachelson, A. (2017). *The Seven Competences of the Sustainable Professional.* Abingdon: Routledge - Greenleaf.

[21] Roorda, N. (2016). *The Seven Competences of a Sustainable Professional. The RESFIA+D Model for HRM, Education and Training* in *Management for Sustainable Development*. Aalborg: River Publishers, 1–48.

[22] Roorda, N. *RESFIA+D*. Available at: https://niko.roorda.nu/management-methods/resfia-d/

[23] Deming, W. E. (1986). *Out of the Crisis*. Cambridge, MA: MIT Press.

[24] Fortune. (2017). *Fortune Global 500, Fortune, 2017*. Available at: http://fortune.com/global500 [accessed June 2017].

[25] Royal Dutch Shell. (1991). *Climate of Concern, Royal Dutch Shell*. Available at: https://www.youtube.com/watch?v=0VOWi8oVXmo [accessed 2017].

[26] Carrington, D. and Mommers, J. (2017). *Shell knew: Oil Giant's 1991 Film Warned of Climate Change Danger, The Guardian*. Available at: https://www.theguardian.com/environment/2017/feb/28/shell-knew-oil-giants-1991-film-warned-climate-change-danger [accessed June 2017].

[27] Royal Dutch Shell PLC. (1998). *Profits and Principles—does there have to be a Choice? The Shell Report 1998*. Available at: reports.shell.com/sustainability-report/2012/servicepages/previous/files/shell_report_1998.pdf [accessed 2017].

[28] Mulvey, K. and Shulman, S. (2015). *The Climate Deception Dossiers*. Available at: http://www.ucsusa.org/sites/default/files/attach/2015/07/The-Climate-Deception-Dossiers.pdf [accessed 2017].

[29] Follow This. (2017). *Motivate Shell to go Green*. Available at: https://follow-this.org/en/ [accessed 2017].

[30] Royal Dutch Shell PLC. (2016). *Notice of Annual General Meeting*. The Hague: Royal Dutch Shell PLC.

[31] Royal Dutch Shell PLC. (2017). *Annual General Meeting, Royal Dutch Shell PLC*. Available at: http://www.shell.com/investors/retail-shareholder-information/annual-general-meeting.html [accessed 2017].

[32] Royal Dutch Shell PLC. (2017). *Notice of Annual General Meeting*. The Hague: Royal Dutch Shell PLC.

[33] Lazard. (2016). *Levelized Cost of Energy Analysis 10.0, Lazard investment bank*. Available at: https://www.lazard.com/perspective/levelized-cost-of-energy-analysis-100/ [accessed 2017].

[34] Bridges, W. (2000). *The Character of Organizations: Using Personality Type in Organization Development*. Palo Alto, CA: Davies-Black Publishing.

[35] Briggs-Myers, I. (1980). *Gifts Differing: Understanding Personality Type*. Palo Alto, CA: Davies-Black Publishing.

[36] Roorda, N. (2010). *Sailing on the Winds of Change. The Odyssey to Sustainability of the Universities of Applied Sciences in the Netherlands*. Maastricht: Maastricht University.

[37] Europe-Third World Centre (CETIM); Env. Rights Action (ERA); Friends of the Earth Nigeria (FoEN) (2014). *Cases of Environmental Human Rights Violations by Shell in Nigeria's Niger Delta. Joint written statement to the Human Rights Council, Twenty-sixth session, Agenda item 3, A/HRC/26/NGO/100,"* 26 May 2014. Available at: http://www.cetim.ch/cases-of-environmental-human-rights-violations-by-shell-in-nigeria%E2%80%99s-niger-delta/ [accessed 2017].

[38] van Beurden, B. (2017). *Spokesman of Royal Dutch Shell, Interviewee, Private E-mail to the Author.* http://royaldutchshellgroup.com/category/ben-van-beurden/

[39] GRI (2017). *Global Reporting Initiative*. Available at: https://www.globalreporting.org/Pages/default.aspx [accessed 2017].

[40] ISO (2017). *ISO 26000 – Social Responsibility*. Available at: https://www.iso.org/iso-26000-social-responsibility.html [accessed 2017].

[41] Roorda, N. (2017). *FFEA Extensions*. Available at: https://niko.roorda.nu/management-methods/ffea-extensions/

[42] Roorda, N. *STELES – Self-Test Leadership Styles*. Available at: https://niko.roorda.nu/management-methods/ffea-extensions/self-test-leadership-styles/

[43] Roorda, N. *FFEA Certificate*. Available at: https://niko.roorda.nu/management-methods/ffea-details/#Certification

[44] Roorda, N. *Book: The Seven Competences*. Available at: https://niko.roorda.nu/books/the-seven-competences/

[45] Roorda, N. *RESFIA+D*. Available at: https://niko.roorda.nu/management-methods/resfia-d/

[46] Roorda, N. *Book: Fundamentals of Sustainable Development*. Available at: https://niko.roorda.nu/books/fundamentals-of-sustainable-development/

[47] Roorda, N. *Workshop: In the Rucksack*. Available at: https://niko.roorda.nu/management-methods/ffea-extensions/in-rucksack/

[48] Roorda, N. *Serious Game: EPU, Environmental Pollution Unit.* Available at: https://niko.roorda.nu/computer-programs/epu-environmental-serious-game/

[49] Roorda, N. *Seriou Game: PopSim, Population Simulation.* Available at: https://niko.roorda.nu/computer-programs/popsim-population-simulation/

[50] Roorda, N. *Forum: Function Analysis.* Available at: https://niko.roorda.nu/management-methods/ffea-extensions/function-analysis/

[51] Roorda, N. *Forum: Backcasting the Future.* Available at: https://niko.roorda.nu/management-methods/ffea-extensions/backcasting-future/

[52] Roorda, N. *Assessment: CSR Motivation Mix.* Available at: https://niko.roorda.nu/management-methods/ffea-extensions/csr-motivation-mix/

[53] Roorda, N. *Workshop: The Societal Triangle.* Available at: https://niko.roorda.nu/management-methods/ffea-extensions/societal-triangle/

[54] Roorda, N. *Forum: The Pledge.* Available at: https://niko.roorda.nu/pledge/

[55] Roorda, N., Rammel, C., Waara, S. and Fra Paleo, U. (2009). *AISHE 2.0 Manual: Assessment Instrument for Sustainability in Higher Education.* Available at: https://www.box.net/s/0dglhugzyyzta4kkfb83

[56] De Caluwé, L. and Vermaak, H. (2006). *Leren veranderen. Een Handboek Voor De Veranderkundige*, 2nd Edn. Deventer: Kluwer.

[57] Taylor, F. (1947). *Scientific Management.* New York, NY: Harper and Row.

[58] Sterling, S. (2004). "Higher education, sustainability and the role of systemic learning," in *Higher Education and the Challenge of Sustainability* (Dordrecht: Kluwer), 47–70.

[59] ISO *ISO 9000 – Quality Management, International Organization for Standardization.* Available at: https://www.iso.org/iso-9001-quality-management.html

[60] EFQM. (2017). *Leading Excellence, European Foundation for Quality Management.* Available at: http://www.efqm.org/ [accessed 2017].

[61] Van Nuland, Y. Broux, G. Crets, L. De Cleyn, W. Legrand, J. Majoor, G. and Vleminckx, G. (1999). *Excellent. A guide for the Implementation of the Efqm-Excellence Model.* Belgium: Balnden.

[62] INK. (2017). *Kwaliteitsmanagement INK.* Available at: http://www.ink.nl/ [accessed 2017].

[63] INK (2000). *Gids Voor Toepassing Van Het INK-Managementmodel.* 's-Hertogenbosch: INK.

[64] Expertgroep HBO (1999). *Method for Improving the Quality of Higher Education Based on the EFQM Model.* Groningen: Hanzehogeschool.

[65] van Kemenade, E. A. (2004). *Methode Voor Kwaliteitsverbetering Van Het Hoger Onderwijs op Basis Van Het EFQM-Model.* Groningen: Expertgroep HBO.

[66] van Kemenade, E. A. (2009). *Certificering, Accreditatie en de Professional. Case Study over Hogescholen.* PhD Thesis, Rotterdam: Erasmus University.

[67] van Kemenade, E. A. and van Schaik, M. (2006). "Interne kwaliteitszorg: van ambacht naar visie," in *Vernieuwing in het hoger onderwijs – Onderwijskundig Handboek* (Assen: Van Gorcum)

[68] IEEE (1992). *Standard: BSI – BS 7750: Environmental Management Systems.* Available at: http://standards.globalspec.com/std/821210/bsi-bs-7750 [accessed 2017].

[69] European Commission. (1993). *EMAS – Environmental Management Systems Council Regulation 1836/93.* Brussels: European Commission.

[70] ISO (2017). *ISO 14000 family - Environmental Management, International Organization for Standardization.* Available at: https://www.iso.org/iso-14001-environmental-management.html [accessed 2017].

[71] ISO (2017). *Iso 26000 – Social Responsibility, International Organization For Standardization.* Available at: https://www.iso.org/iso-26000-social-responsibility.html [accessed 2017].

[72] GRI *GRI Standards, Global Reporting Initiative.* Available at: https://www.globalreporting.org/Pages/default.aspx

[73] Roorda, N. (2016). *AISHE, Assessment Instrument for Sustainability in Higher Education, Niko Roordai/Stichting Duurzaam Hoger Onderwijs.* Available at: https://niko.roorda.nu/management-methods/aishe/ [accessed 2017].

[74] Roorda, N. (2001). *AISHE – Assessment Instrument for Sustainability in Higher Education.* Amsterdam: Stichting Duurzaam Hoger Onderwijs.

[75] Roorda, N. (2004). "Policy development for sustainability in higher education – results of AISHE audits," in *Higher Education and the Challenge of Sustainability* (Dordrecht: Kluwer), 305–318.

[76] Brandli, L. Frandoloso, M. Roorda, N. Fraga, K. and Vieira, L. (2014). Evaluation of sustainability using the AISHE instrument: case study in a Brazilian University. *Braz. J. Sci. Technol.* 1, 1–13.

[77] Togo, M. and Lotz-Sisitka, H. (2014). "The unit-based sustainability assessment tool and its use in the UNEP mainstreaming environment and sustainability in African Universities partnership," in *Sustainability*

Assessment Tools in Higher Education Institutions (Berlin: Springer), 259–288.

[78] Horton-Williams, C. (2010). *Higher Education Leadership Stages and Strategies that Relate to Campus Environmental Sustainability as U.S. Colleges and Universities*, Ann Arbor, MI: ProQuest Dissertations Publishing.

[79] Alshuwaikhat, H. Adenle, Y. and Saghir, B. (2016). Sustainability Assessment of Higher Education. *Sustainability* 8:750.

[80] Shamsuddoha, M. Nasir T. and Gihar, S. (2010). *Sustainable Education and Bangladesh*. Available at: https://papers.ssrn.com/sol3/papers.cfm?abstract_id=1543295 [accessed 2017].

[81] Schultz, M. (2013). Embedding environmental sustainability in the undergraduate chemistry curriculum: a case study. *J. Learn. Des.* 6, 20–33.

[82] International Platform for Sustainability Performance in Education EAUC (2017). Available at: http://www.eauc.org.uk/theplatform/home [accessed 2017].

[83] The Tools of the Platform for Sustainability Performance in Education EAUC (2017). Available at: http://www.eauc.org.uk/theplatform/getting_started [accessed 2017].

[84] TCA (2011). *Auditing Instrument for Food Security in Higher Education, Technical Centre for Agricultural and Rural Cooperation*. Available at: http://aifshe.cta.int/en/

[85] Bello, O. (2013). *AIFSHE – Auditing Instrument for Food Security in Higher Education*. Wageningen: Wageningen University.

[86] Roorda, N. (2014). "A strategy and a toolkit to realize system integration of sustainable development (SISD)," in *Sustainability Assessment Tools in Higher Education Institutions: Mapping Trends and Good Practices Around the World* (Berlin: Springer), 101–120.

[87] NVAO erkent duurzame ontwikkeling als 'bijzonder kenmerk' bij accreditatie (2007). Available at: https://www.denederlandsegrondwet.nl/9353000/1/j9vvihlf299q0sr/vhs7nseheuyw?ctx=vhhfh58d8uzm&start_tab0=80 [accessed 2017].

[88] Nederlands-Vlaamse Accreditatie Organisatie [NVAO] (2017). Available at: https://www.nvao.net/ [accessed 2017].

3

Corporate Social Responsibility: The Case of East Timor Multinationals

Carla Freire, Manuel Brito and Iris Barbosa

Department of Management, School of Economics and Management,
University of Minho, Braga, Portugal

Abstract

This chapter aims to analyze the perceptions of Foreign Multinational Company (MNC) managers with corporate social responsibility (CSR) responsibilities in East Timor companies. Since this country is rather new, and is thus addressing complex social and economic problems, the pertinence of CSR by companies with more resources and knowledge is even greater. This study is based on the results of in-depth interviews conducted with 10 MNC managers who are responsible for CSR issues. The results point to specific social, economic, and cultural reasons in the definition of stakeholders and their relative importance to CSR. The shareholders and employees comprise the group of internal stakeholders, who are referred to most often; the external stakeholders consist of the Government and public institutions, NGOs, as well as religious and educational institutions. Society plays an extremely relevant role in the survival/prosperity of Timor Leste companies. The population's social and economic needs also exercise an influence on the definition of CSR measures. Environmental concerns are less obvious and are considered only when there is a negative environmental impact ensuing from the company's activities. The Government is assessed as being an "external priority" stakeholder due to the roles it accumulates: besides being a regulator, it is also a shareholder and main customer of the companies. Ambiguities in Government regulations, as well as bureaucratic issues, compounded by logistic problems, result in difficult access to the important resources required to respond to stakeholders' requests. These aspects are considered to constitute the main obstacles to the development and implementation of CSR measures in East Timor companies. Future research should focus on CSR practices in developing countries as East Timor to gain a better understanding of the guidelines issued by headquarters and their adjustment to local characteristics.

3.1 Introduction

Before it became popular in the Anglo-Saxonic context, and later in Europe, the concept of *social responsibility* was the object of Howard Bowen's study in 1953. In his book, *"Social Responsibilities of The Businessman,"* he highlighted the businessman's responsibilities concerning to society [1]. The concept has taken on different meanings since then, and became popular in the 1990s [2]. Although CSR has been supported by academics and professionals, its process of development has also met with some resistance by several academics such as Milton Friedman and Christina Keinert. Friedman [3] claimed that the company exists only to maximize profits for its shareholders. This author therefore considered the involvement of companies in social activities to be bad management practice. Keinert [4] reinforced the idea that managers have one single duty: that of maximizing profits for the shareholders, since these possess a work contract with the agents who represent the company's proprietors. This view argues that the company has one sole responsibility – profit; the other issues (social and environmental) constitute a responsibility which should be attributed to the Government and social institutions. Despite the criticism presented by these perspectives, CSR has progressed from ideology to reality [5], and has been implemented worldwide.

Corporate social responsibility (CSR) currently constitutes an important theme and has contributed significantly to the improvement of social, economic, and environmental issues. The concept has, however, developed along the lines of different perspectives [6], and has resorted to varied terminologies [7, 8]. As such, it is associated to complementary concepts, which are sometimes contradictory [9]. Consequently, it has been difficult to reach an agreement on the paradigm or model to follow when studying this theme [8]. As has been stated by some authors, CSR is perceived differently in different countries or locations [10–13], at different moments in time and by different authors [14–17].

Matten and Moon [18] pointed out that not only are there divergences in the meaning of CSR, but its ensuing practice also varies from country to country. Across the globe, companies present differences in the principles, policies, and practices of social responsibility [19, 20], operating through different means of action in different parts of the world [21–23].

These divergent applications of the concept stem from differences in social and cultural contexts, as well as from the distinct values and political history [14, 24, 25] of the organizational context itself. Examples of these are corporate culture [26, 27], financial resources, and experience in CSR implementation, policies, strategies, the size of the company [28], as well as the level of relations of the company with headquarters or any other company it may depend on [29, 30].

Literature thus provides indications as to the plurality of the comprehension of this concept and its varied implementations worldwide. In this sense, and despite the importance of the issue, there are rather few studies which seek to gain a better understanding of the application of the concept in different contexts [31–38]. Existing studies highlight CSR practices developed in the region of southern Asia, southeast Asia [39], and in small countries, former colonies, and emerging economies [37, 40]. Greater attention has always been dedicated to developed countries [31, 37, 40, 41], as well as those which present an advanced economy and a stable political context [42, 43]. This is the case of some of the BRIC member states, such as India and China [39].

In spite of the limited number of studies in these countries, some research on CSR practices has been undertaken in China [44–47], India [48–52], Malaysia [53–57], Bangladesh [34, 58–60], Sri Lanka [61], Pakistan [62], Vietnam [63], and Indonesia [64, 65]. Other studies involving various countries in this region have been conducted, namely, the study by Welford and Frost [43] which includes China, Malaysia, Cambodia, Thailand, Vietnam, and Indonesia, as well as the comparative study by Chamber, Chapple, Moon, and Sullivan [22], which involves South Korea, Thailand, Singapore, Malaysia, Philippines, Indonesia, and Japan, comparing these to the United Kingdom.

These studies address specific subject areas, which are presented in Table 3.1.

Although there are already various studies on CSR in developing countries (as described above), there is still a gap in the literature regarding its implementation. This dimension is one of practical interest, but which has been neglected by studies [69]. Aguinis and Glavas [70], Maon et al. [5],

Table 3.1   Studies on corporate social responsibility (CSR) in the context of Asia

Subject Area	Studies
The significance of CSR	Rani and Khan [66]; Hossain et al. [58]; Munasinghe and Malkumari [61]; Musdiana et al. [57]; Abdul and Ibrahim [67]
The reasons for CSR	Fauzi [64]; Yin and Zhang [44]; Hossain et al. [58]; Muwazir [36]; Hossain and Rowe [59]; Wang and Chaudhri [47]; Quazi et al. [60]
The obstacles in CSR	Hossain et al. [58]; Arevalo and Arivind [51]; Belal and Cooper [34]; Zabin [56]
Corporate social responsibility policies and strategies	Pradhan and Ranjan [52]
Implementation of CSR	Pradhan and Ranjan [52]
The impact/benefits of CSR	Trang and Yekini [63]; Chen et al. [68]; Pradhan and Ranjan [52]
The focus of activities and stakeholders	Kulkarni [48]; Rani and Khan [49]; Yam [55]; Musdiana et al. [57]; Asniwaty [65]; Tang and Li [46]

and Husted and Allen [71] have underlined this precise need to observe the implementation process in these countries. This chapter seeks to bridge this gap by presenting the way in which CSR has been implemented in large subsidiary companies and/or affiliates in East Timor. It further attempts to identify the difference between the CSR practiced by companies when compared to their headquarters, and/or the organization they depend on, from the perspective of globally mobile expatriate or local managers.

According to Basu and Palazzo [72] and Maon et al. [69], studies in this field have been aligned to address three main questions: (i) *"what,"* which focuses on a descriptive analysis; (ii) *"why,"* which explores the reasons for and the driving forces behind CSR; and (iii) *"how,"* which essentially deals with the CSR implementation process.

The present study aims to analyze CSR considering the same three domains: perceptions *(thinking)*, implementation *(doing)*, and consequences *(consequence)*. In the domain of *thinking*, the objective resides in analyzing the perceptions and reasons why companies have become involved in CSR, while in the domain of *doing*, one seeks to gain an understanding of CSR strategies and their implementation. Finally, in the domain of *consequence* in CSR, and besides assessing the activities and benefits thereof, one has also evaluated the different forms of CSR practiced by companies in East Timor when compared to their parent companies or the organizations they depend on.

Stated briefly, the objectives of this study aim to answer the following questions:

- How is CSR understood by the managers of Foreign Multinational Company (MNC) affiliates with CSR responsibilities?
- Who are the main *stakeholders* and what are the issues inherent to CSR in East Timor?
- Why do companies become involved in CSR practices and how does this occur?
- What are the benefits/advantages of implementing CSR practices for companies?
- Is the CSR undertaken by the affiliate companies in East Timor different from that of their foreign headquarters?

This study aims to contribute to a greater knowledge of social responsibility in developing countries and emerging economies. The companies in East Timor have been selected as the object of study for several reasons. The first of these is that, at present, there are no CSR studies relating to this country, so that one knows nothing of the nature of CSR in East Timor. Second, the country was both greatly influenced by a history of traditional philanthropy before and after external domination, as well as by the introduction of global practices on the part of trans-governmental organizations and global institutions during the United Nations transition and post-transition periods. Consequently, some of the global corporate practices which have been institutionalized in this country include CSR practices.

3.2 Theoretical Framework: Corporate Social Responsibility Theories

3.2.1 The Stakeholder Theory

The stakeholder theory deals with the relations between companies and society [73], and expands the traditional objective of profitmaking to include a wider scope. Namely, company objectives are no longer considered to reside in the maximization of profits but also in social intervention, thus addressing the interests and demands presented by the different parties involved (shareholders, workers, suppliers, customers, trade unions, NGOs, etc.).

Seen from under this theoretical lens, the concept of CSR has already been used in several studies [7, 74]. The stakeholder theory explains the relationship between the company and its stakeholders [75], addressing the issues of profit redistribution and corporate power [76] with the purpose of balancing business interests with those of the stakeholders' [15].

The sustainability of companies has been interpreted as consisting of the capacity to balance economic objectives with social objectives [77]. The attempt to strike a balance between the business and social dimensions is a theme that has been explained by various authors [76, 78, 79]. They underline the importance of the role played by companies in generating wealth, value, or satisfaction, both for the shareholders involved as well as for the other stakeholders, so that the survival of the business and its development are assured [69, 80].

Several authors have defined the "stakeholders" as comprising the person or group who affects or is affected by the decision-making process, as well as by the company's policy and its performance [76, 78, 81]. Clarkson [82] defines the concept as being the person or group of people who possess ownership rights over or interest in an organization. Öberseder et al. [83] suggest that the stakeholders include the persons, groups, organizations, and institutions, of society and of the environment. Based on this perspective, companies classify and prioritize the different interests and demands presented by these agents in order to meet their expectations [75, 82].

Carroll [84] identifies the stakeholders as: shareholders, employees, consumers, suppliers, the local community, competitors, interest groups (representatives of civil society), the Government, press, and general society. Zhao et al. [85] also identify the stakeholders as being employees, consumers, suppliers and partners, the local community, competitors, NGOs, the Government, shareholders, credit providers, and environmental groups.

Some authors have differentiated the stakeholders and divided them into two groups: the primary and the secondary [69, 82]. The primary group consists of consumers, employees, partners, shareholders, and suppliers; this group is essential to the company's survival due to the interdependent and contractual relations established with the company [82]. The secondary group comprises the Government, non-government organizations, the local community, the press, and the environment. This group is characterized by the inexistence of a formal or contractual relationship with the company and is, therefore, less dependent on it [82, 86]. Maon et al. [69] classify the primary group as being vital to an organization's achievement of the mission to produce goods or a service, and includes consumers, managers and employees, the Government, suppliers, and investors. The secondary group is constituted to support the company's mission, thus enabling its legitimacy.

It consists, above all, of the press and the local community, competitors, and non-profit organizations.

Another classification of the stakeholders is based on the level of proximity. Jones [87] distinguishes two groups of stakeholders: the internal and external. According to the author, the internal stakeholders are the groups or entities which greatly influence the organization, namely, the shareholders, managers, and employees; the external stakeholders are those groups who have interests and thus influence the company externally, such as customers, suppliers, the Government, commercial associations, the local community, and society at large.

Freeman and Liedtka [88] propose another form of stakeholder categorization and base this on the degree of cooperative potential and on threats from the competition. Swing stakeholders are the groups which influence the company positively or negatively through changes in their behavior; defend stakeholders are the groups which already contribute to the company, and their support plays a crucial role in ensuring its performance; opportunity stakeholders are the groups available for the execution of smaller activities to assist the company in reaching specific established objectives; and monitor stakeholders are groups which could be significant; however, when required to make a behavior change, their choices are limited, and so they have little leeway or limited flexibility when influencing the company's performance.

The four stakeholder profiles ensue from the intersection of the degree of cooperative potential and that of competition threats [88], as presented in Figure 3.1

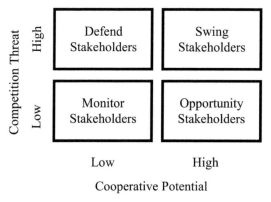

Figure 3.1 Four stakeholder profiles.

Donaldson and Preston [89] subdivide the stakeholder theory into three distinct perspectives: descriptive, instrumental, and normative. The descriptive perspective emphasizes the way in which a company represents and understands the relationship it establishes with the internal and external environments. The instrumental perspective highlights the use of management tools to pursue economic value, focusing on management by the stakeholders. The normative perspective underlines the ways of managing and communicating with all the stakeholders. While the normative perspective explains how the stakeholders should be dealt with, the descriptive approach broaches the ways in which the stakeholders should be managed and communicated with. Finally, the instrumental perspective points to how the stakeholders should be managed in order to meet economic targets.

The three views concerning stakeholder theory are divided into two opposing branches: the moral and the strategic perspectives [89–93]. The stakeholder theory which follows the normative or moral branch is grounded on ethical and moral philosophy. This ethical principle is referred to by Jones [87] as being the moral principles which indicate what is "right" or "wrong," and whose purpose is that of guiding managers in decision-making, especially helping them to solve ethical dilemmas and to treat stakeholders better [87]. The normative dimension of the stakeholder theory emphasizes that the company is a part of society and should thus behave in an ethical manner, paying attention to society's well-being [91, 94], without considering its contributions to the company [95]. In line with this perspective, managers should treat all the stakeholders equally, without taking into account their power and impact on the company's financial performance. According to the point of view of normative theory, the organization should act as a mechanism which treats all stakeholders fairly, so that managers should follow this maxim when solving dilemmas that involve a diversity of stakeholder interests. This view affects the way the company treats some of its stakeholders, which may produce reactions of dissatisfaction with the company.

On the other hand, and in accordance with the strategic approach, the stakeholder theory explicitly refers to the stakeholder issues and forces which influence the organization in such a way that it will adjust its behavior to meet stakeholders' expectations [93]. The stakeholder theory acknowledges that a company will establish a different relationship with each of its stakeholders during its operation, thus addressing the different expectations at hand. However, it should try to balance the stakeholders' different interests [96] and, above all, clearly explain how managers should balance the various stakeholders' conflicting interests to ensure the company's survival in the long term [97]. In this situation, and owing to the stakeholders' diverging interests, managers are confronted with ethical dilemmas when making decisions [98], especially when this concerns the company's capacity and impacts on its economic performance. It is in this sense that companies manage their relationship with stakeholders in a pragmatic manner.

Since it is impossible to meet all the stakeholders' demands, companies tend to focus their attention on the most important groups, thus ensuring the company's existence and sustainability. This pragmatic behavior is influenced by both external and internal factors. The external factors include, for example, the complexity and contradictions presented by the stakeholders, while the internal factors are related to the organization's limited resources. With regard to the strategic perspective, and in accordance with Freeman [75], the company only tries to respond to the powerful groups, namely, those which can benefit the organization, due to their legitimacy and license to operate, risk management, and learning curve. Thus, the company favors the stakeholder group in possession or in control of the resources considered to be the most important for the organization, be they tangible (natural, financial, and human resources) or intangible (political power and/or legitimacy). The stakeholders are therefore defined according to how important they are to the company.

Mitchell et al. [99] proposed the "salience stakeholder" model, which deals with stakeholder identification and priority. It anchors on the stakeholders' degree of influence, namely, their degree of power, legitimacy, and urgency. This mechanism may assist managers in dealing with these parties in a more effective and efficient manner. According to the authors, stakeholders can be identified through three distinct features: (1) power – the degree of stakeholders' power can influence the organization; (2) legitimacy – the level of legitimacy granted to an organization when a relationship is established between the company and its stakeholders; and (3) urgency – which relates to the urgency of the stakeholders' claims on the organization.

The typology presented by Mitchell et al. [99] advocates seven types of stakeholders: (1) the dormant stakeholder; (2) the discretionary stakeholder; (3) the demanding stakeholder; (4) the dominant stakeholder; (5) the dangerous stakeholder; (6) the dependent stakeholder; and (7) the definitive stakeholder (Figure 3.2).

Three classification classes ensue from the "salience stakeholder" model: latent, expectant, and definitive. The latent stakeholders are subdivided into: (1) Dormant – the stakeholders who only possess power. Although these stakeholders cannot be used, they are worthy of attention because they have

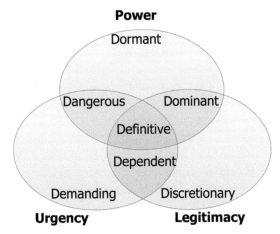

Figure 3.2   The "salience stakeholder" typology of Mitchell et al. [99].

the potential to take on other features; (2) Discretionary – those who only possess legitimacy. Companies can satisfy this group by providing philanthropic action; (3) Demanding – the stakeholders characterized by urgency. These stakeholders must be monitored, especially since they can acquire other characteristics. The next group consists of the expectant stakeholders, who are divided into: (4) Dominant – the stakeholders with both power and legitimacy. These stakeholders must be given special attention; (5) Dependent – the stakeholders who possess the features of legitimacy and urgency. This group must be taken into account by the company because they exercise indirect power; namely, they relate with the powerful stakeholders to influence the company, thus ensuring that their claims will be met; (6) Dangerous – urgency and power are the characteristics associated to these stakeholders. Managers must, therefore, consider them to be a priority as they can coercively influence the company. Finally, definitive stakeholders are those which exercise power, legitimacy, and urgency. It is to these that the company will direct most of its attention and priority.

3.2.2 The Institutional Theory

According to institutional theory, organizations must adapt to the social and cultural expectations imposed by the institutional environment in order to achieve success and survive [100]. This theory emphasizes the issue of context and its importance when determining CSR policies and practices [93, 101, 102]. Institutional theory has been harnessed to gain a better understanding of social responsibility and of the issues related to stakeholders [87, 103], organizational legitimacy [104], and the wider implementation of CSR [18].

DiMaggio and Powell [105] and Lee [104] agree that organizations adapt to the institutional environment with the purpose of achieving legitimacy and social conformity. Once this legitimacy has been assured, it provides other benefits, such as the possibility of obtaining operation licenses and having access to the resources available in the location where the firm has been installed. Social conformity and legitimacy are achieved in an interconnected and sequential manner; namely, legitimacy can be maintained or improved if the organization conforms socially, especially if it is able to adapt to the values, norms, and beliefs of the society in question. On the other hand, if the organization does not adapt to the values and institutional norms at hand, its legitimacy may weaken and it might even place its survival at risk [105].

For Scott [106], the three pillars of institutional order are: regulative, normative, and cognitive. The regulative pillar concerns rules, monitoring, and sanctions. The normative aspect is related to compliance with ethical duties and social norms. Finally, the cultural-cognitive pillar is based on the perception of symbols, words, and signs which produce an effect on the way one builds the meaning of objects [106]. An organization may be legitimized by society if it is able to adjust to the institutional aspects of that society. This process could result in behavioral homogeneity, known as the phenomenon of isomorphism [87, 105]. DiMaggio and Powell put forward the three main types of institutional isomorphism: (a) coercive isomorphism, when the organization adopts specific values and norms due to the pressure from external factors, namely, those which ensue from public organizations (local and global), as well as from the organizations it defends (for example, its business partner, the Government, consumers, and NGOs). The parent company or partner thus spurs the adoption of CSR practices by dependent companies, for instance, through the establishment of a code of ethics or guidelines directed at managers, so that conflicts with local stakeholders are minimized [107]; (b) mimetic isomorphism occurs when an organization imitates the practices of other organizations to increase its legitimacy and thus meet the uncertainty of the organizational environment; and (c) normative isomorphism, in which an organization is voluntarily similar to another, as a result of professional demand or as a consequence of its affiliation with another organization. The author further explains that this normative isomorphism occurs through formal education (institutions of higher education) and non-formal education (training).

Various empirical studies have highlighted the influence of institutional mechanisms in the adoption of CSR practices by companies, both in developed countries [101] and in developing ones or those with emerging economies [35, 44, 108]. Other studies have additionally underlined the importance of the influence exercised by institutions on the socially responsible behavior of companies [15, 109]. In many cases, adaptation to these practices is influenced by factors which are external to the organization [79], whether in an individual or organizational form, by national or transnational agents [110].

There are two types of pressures which affect the adoption of CSR practices: the macro and micro contexts. At the macro level, the pressure stems from regulating entities and global business partners [44], while the micro level is related to pressures which ensue from the host country. As such, expected behavior must comply with the beliefs and values at hand in order to obtaining and maintain external legitimacy [105, 111].

Campbell [15] provides a more detailed explanation of the influence of the local institutional framework on the socially responsible behavior of companies. In order for companies to be socially responsible when operating in conditions of significant regulation, self-regulation must be implemented. This control must be undertaken in associations linked to industry, in employment institutions, and in non-profit organizations. A normative institutional environment must be ensured to enable dialog amongst stakeholders. Jamali and Mirshak [37] and Matten and Moon [18] also point to a greater difficulty in implementing CSR in some countries due to their levels of institutionalization. An example of this is the study by Sánchez-Fernández [101], which was undertaken in the region of the Iberian Peninsula, more specifically in the south of Spain and in the north of Portugal. It concluded that the coercive and normative institutional mechanisms involved determined the existence of CSR practices. This coercive isomorphism includes, for instance, a knowledge of legislation, the application of laws, and the existence of agreements. Normative isomorphism pertains to moral duties, adaptation to the context and to the organizational environment, as well as adjustment to social norms.

In their study, Fifka and Pobizhan [108] analyzed CSR from an institutional perspective, involving 50 large Russian companies. The authors categorized CSR practices in Russia and divided these into two institutional models: the first was influenced by the country's political, social, and economic environments, while the second was exposed to the influence of a global institutional environment. In order to improve a company's legitimacy, these authors underline the importance of meeting global demands to conform with international norms.

The study carried out in China by Yin and Zhang [44] is also related to institutional perspective of CSR, but places greater focus on the sequential order of the mechanism of isomorphism. During the first phase, the adaptation of CSR is caused by the mechanism of coercive isomorphism exercised by the regulating entities and global business partners. Subsequently, the normative mechanism plays its role in CSR adaptation by means of the association of industry and the dissemination of CSR. This study warns of the phenomenon of the dissemination of CSR from developed countries to developing countries, or emerging economies. In the initial phase, CSR is still not very popular and, as such, companies are coercively guided by global agencies. During a later phase, and after pioneer companies have been institutionalized, CSR begins to extend to other firms through the diffusion and training enabled by industrial associations and educational institutions.

A study undertaken in Kenya by Muthuri and Gilbert [35] demonstrates that in a country where few regulatory pressures are exercised by the Government, the process of CSR adaptation is driven by the need to conform to social norms and global practices to gain legitimacy and a competitive advantage. In this context of economic globalization, and in countries where the regulatory systems are weaker, companies imitate the best global company practices. Simultaneously, they align with their organizational values in order to increase their competitive edge.

3.2.3 The Theory of Legitimacy

The theory of legitimacy indicates that this aspect represents an important intangible resource for companies, one which is granted by the community [112]. Not only does it legitimize its existence, but it also permits its survival.

For Suchman ([103], p. 574), legitimacy means "a generalized perception or assumption that the actions of an entity are desirable or appropriate within some socially constructed system of norms, values, beliefs, and definitions." Scott [113] highlights four elements that enable legitimacy: regulations, norms, as well as the culture-cognitive and moral aspects. Regulative legitimacy ensues from the regulations, rules, patterns, and interests created by governments, regulating agencies, professional associations, and influential organizations [114]. This legitimacy possesses formal and coercive characteristics. The legitimacy which derives from norms demands compliance with these norms, as well as with the informal values accepted by society. The organization is welcomed when it behaves in accordance with the society's values, namely, in tune with the social environment in which it operates. Furthermore, legitimacy – which anchors on cultural-cognitive aspects – is related to the compatibility of organizational values with cultural beliefs, which are extensive and are still practiced by the community [113]. Finally, moral legitimacy consists of a choice made in accordance with the behavioral practices of what is considered to be "right" or "wrong" [103].

Legitimacy thus constitutes a key factor for companies [115], and is an important resource to ensure sustainability and development [105, 113, 116, 117]. This is achieved through efforts to align organizational behavior with the society's values and norms [93, 117] and by addressing its expectations [118]. An organization is able to achieve legitimacy when there is compatibility between company practice and institutional regulations, norms, and social expectations [119–121]. By subsequently using this legitimacy, companies can then gain access to the community's resources, as well as obtain support for their operations.

Suchman [103] also attempted to deal with the subject of legitimacy, stating that it can be achieved indirectly, through intermediate organizations. In this case, the company gains society's legitimacy through other organizations which connect the company with society.

Legitimacy possesses dynamic features. Namely, the company's values may sometimes not conform to those of society and thus trigger a negative public reaction [122], which might represent serious consequences for the company [123]. Therefore, the legitimacy theory advocates that organizations should constantly evaluate the compatibility of their behaviors and values with society's norms [124], especially with regard to policies, decisions, and actions. By acting in accordance with society's expectations, this condition will allow for the company's survival [125]. The *"legitimacy gap"* occurs when organizational performance is not compatible with society's expectations and, above all, with that of its stakeholders [126, 127].

The prioritization of stakeholders can be likened to a contract between the community and a company in which, by means of some contributions, the company is able to gain its legitimacy. Namely, it can obtain a license to operate [128], and this can be maintained if the company constantly responds to the demands expressed by the society and stakeholders concerned. These demands chiefly relate to social, ethical, and environmental issues [129, 130], obviously through adaptation and active involvement in issues related to CSR.

According to Palazzo and Richter [131], the acceptance of the social values at hand motivates companies to carry out various initiatives: philanthropy, collaboration with stakeholders, self-regulation, and the dissemination of CSR. A company's legitimacy and reputation can be reinforced by CSR, so that it simultaneously meets the stakeholders' needs and operates lucratively [9]. Along the same lines of thought, Wei et al. [132] point to the fact that those companies which adopt CSR practices in China, especially when these actions are linked to the environment, can enhance their legitimacy and, in turn, see a positive impact on organizational performance. The company's behavior is, however, determined by the legal framework of the countries in question. In their studies concerning the African continent, Amaeshi et al. [133] found that companies in Nigeria use CSR to achieve political legitimacy and to bridge the gap created by the dysfunction of the legal framework. In addition, in a study conducted in 16 Chinese companies, Yin and Zhan [44] concluded that the purpose of CSR practices in this context was identical. In sum, according to this perspective, companies use CSR as a kind of social contract, so as to present themselves as good citizens in the face of society. CSR is not only regarded as a form of obtaining legitimacy but it is also a way of surpassing situations of legitimacy crisis.

3.2.4 Multiple Approaches

The legitimacy theory explains the existence of the company in society; namely, the company constitutes a part of society, so that it must interact therewith. Nevertheless, this theory views society holistically, as an entity which can legitimize or accept the company's existence when its behavior complies with that society's norms, values, and beliefs. In practical terms, these attempts are undertaken through the company's implementation of CSR policies and strategies with a view to ensuring its existence, acceptance, and operation licenses. The application of this theory is questionable: "In a society consisting of different groups, is it possible that these have similar interests, and are they fully addressed?" The stakeholder theory aims to establish which groups in society should be considered by the company, according to its capacity. In this theory, companies favor the groups that influence/benefit the company or which are influenced/jeopardized by it.

Although both theories explain the relationship between company and society, especially with regard to stakeholders, they do not describe the mechanisms (means) to establish this relationship. The role of institutional theory is, therefore, that of explaining how a company can maintain its relationship with society, especially with that of its stakeholders. This theory explains that, in order to establish a good relationship with society, and especially with its stakeholders, the company must be in tune with the institutional environment. This must occur in the location itself, through adaptation to the best practices (adaptation to global patterns and local social values), as well as in a wider global context. The company can then subsequently generate economic and social value, both for itself and for the stakeholders as well.

The three theories contribute to an understanding of the relationship between companies and society. While the legitimacy and stakeholder theories focus on this relationship, institutional theory emphasizes the alignment of company behavior with the expectations of the society and institutional environment it operates in. In sum, the company interacts with society by becoming involved in CSR activities. This can be seen as a type of commitment aimed at establishing good relations with society, especially with regard to the stakeholders, by means of an alignment of organizational values with the existing institutional environment, both in the local and global contexts. The company thus ultimately acquires the legitimacy to ensure its existence and sustainability.

3.3 Methodology

3.3.1 Procedure and Description of the Data Collection Instrument

Semi-structured in-person interviews were conducted with multinational company managers, especially those in Management positions and dealing with CSR or exercising any other position associated with this area in East Timor. The interview guide was drawn up in the Portuguese language and translated into English, Indonesian, and Tetum. The English and Portuguese versions were revised by proficient professionals of these languages.

The interview guide was divided in two sections. The first presented questions which aimed to characterize the interviewee. The second section consisted of questions pertaining to CSR practices: (1) description of job position; (2) perception of CSR; (3) CSR policies and strategies; (4) motivation; (5) CSR management; and (6) the relationship between the company and its headquarters.

A preliminary study was carried out, which involved analyzing the company websites as well as some newspapers. One was able to obtain a database for the TradeInvest East Timor companies (the East Timor public investment agency) and identified 16 companies: three agricultural companies, three banking companies, two civil construction firms, three companies involved in the exploration and distribution of petroleum, two logistics and storage companies, and three telecommunications companies. These 16 companies were considered to be the analysis unit for the definition of the study's sample. The selection of these companies was grounded on the following criteria: they had to be large and linked to foreign investment, with subsidiary or local companies depending on foreign firms. They also had to have been established in East Timor for a period of over 2 years, employing more than 50 staff members and showing involvement in CSR practices.

During the next phase, one contacted the companies directly, and initiated fieldwork in East Timor at the beginning of June 2015. This contact process implied traveling to the country to visit the locations of companies, especially those which had no website or, even though they did, they had not provided us with a complete address. In East Timor, it is essential to travel to companies' locations since the Timorese street coding system, as well as the numbering of buildings and houses, is still rather incomplete. Besides this, another reason is that this ensures direct contact with the companies studied; one can then explore information about the type of company involved and establish a relationship of trust, which will enable one to obtain authorization for the study to be undertaken.

The following phase occurred at the end of June 2015 and consisted of the formalization of a request for authorization to carry out the study. This request consisted of a letter addressed to the head manager at each of the companies. The letter provided a detailed description of the research theme, its objectives, and the importance of the study in question. This document also explained the method used in the investigation and provided assurance of the confidentiality of information and the anonymity of the companies and participants involved.

A total of 16 companies were contacted. Of these, nine authorized the study and the remaining seven were not considered for several reasons: either due to the complexity of the headquarters' authorization procedure, or the managers' availability during the period required for holding the interviews, or the absence of the head manager owing to holiday leave. The nine companies included in the study were: two companies in the area of agricultural business, a banking company, a construction company, two companies associated to the exploration and distribution of petroleum, one logistics and Storage Company, and two telecommunications companies. The managers indicated by companies to participate in the study were from the CSR department or from a similar department, thus ensuring that the participants were responsible for the area of social responsibility at the companies.

The subsequent phase consisted of the moment when we established contact with, or were contacted by, the interviewees to present the study theme more explicitly and to define details pertaining to the study, namely, the interview schedule. Some of the interviews could not take place on the appointed day and had to be rescheduled.

The interviews occurred between the 8th of July and the 5th of October 2015. Eight interviews took place in the municipality of Dili and one interview was held in the municipality of Ermera, which is situated at a distance of 30 km from Dili. The interviews followed three phases: the introductory, development, and final phases. Interviews were initiated with a brief presentation by the interviewer, who also described his/her institution and provided more information about the research theme and its objectives. Further details were also provided for the interview procedure: its approximate time of duration, the guarantee of the interviewee's anonymity and data confidentiality, and the interviewee's right not to answer questions. According to some authors, this phase is extremely important, since the relationship built by both parties – the interviewer and interviewee – is essential to the success of the interviews [134]. This resulted in the strict application of this phase of investigation, thus ensuring that the ethical requirements were met. One sought to build an informal relationship with the interviewees with the purpose of generating a relaxed and trustful environment for both parties involved. This procedure would assure both an interactive dialog as well as a greater reliability of the information provided. After this initial intervention, one requested authorization to record the interview. This was undertaken by means of a signature on a consent form to permit a voice recording, which would also guarantee the confidentiality of the information provided, as well as compliance with the rules established.

The interviews took place at the respective companies, lasted an average time of 1 h, and were recorded. Five interviews were carried out in the Tetum language (the official Timorese language), two interviews were held in the Indonesian language, and the remaining two were carried out in Portuguese and English. During the final interview phase, we gave the interviewee the opportunity to add any information pertaining to what had not been sufficiently explored during the course of the interview.

3.3.2 Sample Description

The qualitative method was selected due to the fact that it allows one to obtain substantial information about the phenomenon under study [135–137], with a greater focus on the semi-structured interview technique. Ten managers from nine Timorese companies were interviewed. The age range of most of the interviewees was between 35 and 54 years (60%). Half of these subjects were of Timorese nationality, while the rest were foreign: 30% were Indonesian, 10% were Portuguese, and 10% were from Singapore. Most of the interviewees were, therefore, from an Asiatic region. Only one of the interviewees was of the female gender. Nine of the subjects possessed Honors degrees, while others also had a Master's degree. Most of the subjects interviewed were intermediate level managers and all of them simultaneously worked in the CSR area. Seven of the 10 interviewees had occupied their job position at the time for less than 3 years.

Table 3.2 Summary of interviewees and companies' characteristics

Company	Origin	Activity Sector	Interviewee	Characteristics
Company 1	Foreign (France)	Logistics and Shipping	Manager 1	Age: 27 years
				Nationality: Timorese
				Gender: Male
				Education level: Master's degree
				Position: HR Director and quality control
				Position tenure: 3 years
				Interview undertaken on: 8/7/2015
Company 2	Foreign (Indonesia)	Telecommunications Service Provider	Manager 2	Age: 54 years
				Nationality: Indonesian
				Gender: Male
				Education level: Honors degree
				Position: Vice President Corporate Secretary
				Position tenure: 2 years

(Continued)

Table 3.2   Continued

Company	Origin	Activity Sector	Interviewee	Characteristics
				Interview undertaken on: 14/7/2015
Company 3	Foreign (Indonesia)	Banking	Manager 3	Age: 49 years
				Nationality Indonesian
				Gender: Male
				Education level: Honors degree
				Position: Operations Director
				Position tenure: 1.5 years
				Interview undertaken on: 16/7/2017
Company 4	Local affiliate to International Cooperative of USA	Agrobusiness	Manager 4	Age: 41 years
				Nationality: Timorese
				Gender: Female
				Education level: Bachelor's degree
				Position: Head of Health division
				Position tenure: 12 years
				Interview undertaken on: 4/8/2015

Table 3.2 Continued

Company	Origin	Activity Sector	Interviewee	Characteristics
Company 5	Local affiliate to a Brazilian multinational telecommunications company	Telecommunications service provider	Manager 5	Age: 55 years
				Nationality: Timorese
				Gender: Male
				Education level: Master's degree
				Position: Director of Institutional Relations and Legal Counsel
				Position tenure: 5 years
				Interview undertaken on: 14/9/2015
Company 6	Foreign (Australia)	Petroleum and Energy Distribution	Manager 6	Age: 35 years
				Nationality: Timorese
				Gender: Male
				Education level: Master's degree
				Position: Local Representative and Community Relations Assessor
				Position tenure: 2 years
				Interview undertaken on: 30/9/2015

(*Continued*)

Table 3.2 Continued

Company	Origin	Activity Sector	Interviewee	Characteristics
Company 7	Foreign (Singapore)	Agrobusiness	Manager 7	Age: 56 years
				Nationality: Timorese
				Gender: Male
				Education level: Honors degree
				Position: Executive Director
				Position tenure: 10 years
				Interview undertaken on: 14/8/2015
			Manager 7a	Age: 51 years
				Nationality: Singaporean
				Gender: Male
				Education level: Honors degree
				Position: CEO
				Position tenure: 6 months
				Interview undertaken on: 14/8/2017
Company 8 (Company group)	Foreign (Portugal)	Construction, Wholesales and Distribution	Manager 8	Age: 42 years
				Nationality: Portuguese
				Gender: Male
				Education level: Master's degree
				Position: CEO
				Position tenure: 2 years
				Interview undertaken on: 5/10/2017

Table 3.2 Continued

Company	Origin	Activity Sector	Interviewee	Characteristics
Company 9	Foreign (Indonesia)	Petroleum and Energy Distribution	Manager 9	Age: 40 years
				Nationality: Indonesian
				Gender: Male
				Education level: Honors degree
				Position: Marketing Director East Timor Representative
				Position tenure: 2 years
				Interview undertaken on: 10/9/2017

Note: Manager 7a accompanied the interviewee Manager 7, with the purpose of reinforcing the statements made by the main interviewee (Manager 7).

3.4 Results

The literature emphasizes the persistent lack of consensus on the general definition of CSR [8, 9]. However, in the specific context of East Timor, and based on the in-depth interviews undertaken with 10 managers performing roles in this area for affiliates of foreign MNCs, various common denominators clearly emerged with regard to CSR definitions and their implications.

The reference to stakeholders was particularly evident in these managers' discourses. They also revealed the perception that CSR policies and practices benefitted both these groups and/or individuals, as well as the companies themselves, thus pointing to a "win–win" situation [69]. Moral and philanthropic reasons were interspersed and balanced by reasons of a more strategic profile, which were used as a justification for the growing concern and involvement of these companies in CSR initiatives.

These issues are explored in the following sub-chapters.

3.4.1 Identifying the Stakeholders

The influence of local socio-economic and cultural conditions on the definition of stakeholders has been consistently identified and emphasized in various national contexts [14, 24]. The managers interviewed in this study of the Timorese context highlighted several stakeholders, including the Government and public institutions, NGOs, consumers/customers, business partners, shareholders, and employees, as well as the community in general. Socio-economic and cultural reasons contribute both to this nomination and to the relative importance of the various stakeholders. For instance, although the Government is often considered by western entities to be a secondary player [82, 138], the discourse of the interviewed managers suggests that it is assessed as an "external priority" stakeholder in East Timor or as "dominant," in line with the typology proposed by Mitchell et al. [99]. This aspect is also revealed in the study undertaken by Yin and Jamali [139] in China. This fact ensues from the accumulation of roles exercised by the Government of East Timor: besides being a regulator, it is also a shareholder and main customer in companies. The following excerpts underline these idiosyncrasies:

> *Our relation with the government is good because the State has the funds. The Government is our indisputable and indispensable partner. As well as the National Election Board, the Anti-Corruption Commission [...], and several ministries. (Manager 5, Telecommunications service provider-Company 5)*

> *Very important to us, to our activity, are the various ministries: the Ministry of Telecommunications, the Ministry of Foreign Affairs, the Ministry of Justice and the SEPOPE (State Secretary for Professional Training and Employment). (Manager 2, Telecommunications service provider Company 2)*

Some of the statements also highlight the importance of general society in the company's survival/continuity, so that its needs are reflected significantly in the definition of CSR measures. Regarding this aspect, the interviews undertaken reveal that there is a focus on the response to social problems, which was also found by Fernando et al. [140] in the context of Sri Lanka. Of these CSR initiatives, several managers emphasized the provision of study grants to young students and occasional collaboration in the training of university students, as reported by the following managers:

> *In 2013, if I'm not mistaken, we sent 10 students to UUI [name of the university], in Indonesia, to continue their studies. (Manager 2, Telecommunications service provider-Company 2)*

We also give lectures to university students. My presence is sometimes requested by university students to present lectures on various areas, in the area of computer technology, on education...
(Manager 5, Telecommunications service provider-Company 5)

Environmental concerns are less evident, and these especially emerge when companies acknowledge the negative environmental impact of their activity. Measures are then referred to address the issues of reforestation, water management, waste treatment, and the control of water and soil pollution. The following statement illustrates this point:

At the start of our activity, we experienced many problems like landslides, deterioration ... So that, from that moment onward, we have tried to keep the environment as close as possible to its natural state. In other words, we have tried to minimize our footprint on the natural environment surrounding us. (Manager 7, Agrobusiness-Company 7)

Other external stakeholders identified by managers can be classified as "swing" stakeholders [88]. This group, which incorporates the key factors required for the sustainability of the respective companies, comprises consumers and clients. It is therefore important to address their expectations of the CSR area [8]. The comment expressed by Manager 2 (Telecommunications service provider-Company 2) reveals this concern: "We improved our service due to customers' complaints. When they complain, it's a sign that we have to do something so that the relationship between the customers and the company can improve."

Additional external stakeholders were considered: these consist of various non-profit organizations, both local and international, as well as educational and religious institutions. One should point out that the Catholic religion is predominant in East Timor, and, as such, religious institutions benefit from great legitimacy in society. The following comment exemplifies corporate response to the needs of educational institutions:

We also cooperate with the UUA (name of the university), more specifically with the teacher training institute. We support it to help its teachers. Due to our collaboration, there are currently eight lecturers involved in Master's studies in Education. (Manager 6, Petroleum and Energy Distribution-Company 6)

On the other hand, while literature acknowledges employees to be one of the key internal stakeholders [82, 141], the data pertaining to this study suggest that less priority is attributed to the CSR measures directed at this group. Indeed, the statements gathered point to the mere existence of training actions, which are seen to be of benefit to the employee and his future employability; yet their immediate aim lies in the improvement of quality and quantity in production. These results seem to ensue from the weak *bargaining power* of employees in the context of East Timor, which is associated to the low rate of worker unionization and limited employment in the country [142]. In line with the classification proposal suggested by Freeman and Liedtka [84], this group could be assessed as monitor stakeholders.

3.4.2 Balancing Moral and Economic Motivations in CSR

The managers who participated in this study provided normative, ethical, or moral reasons, as well strategic ones, to explain their growing involvement in CSR initiatives, suggesting that both objectives can easily be conciliated.

The wish to contribute to the well-being and improvement of life quality in society is often associated to managers' moral values, to their ethical conscience, and to their willingness to do "the right thing" (Manager 7, Agrobusiness-Company 7). This is in line with the literature, which associates these aspects to the normative or moral branch of the Stakeholder Theory [87, 91, 94]. The following comments point to a preoccupation with society and an awareness that companies should develop ethical behavior to benefit social well-being:

> *We have defended for years that we cannot distance ourselves from the community in which we operate. By merely employing workers, the company has already assumed a role concerning the community; namely, it allows people to have jobs. But that is only one of the factors involved and it's not enough. So, we try to support activities related to the social, religious, sports, and cultural areas. We always try to focus on actions of a social nature which will produce immediate and visible benefits to the community (...) All this is voluntary and it happens because we feel that we have the duty to respond to the community's needs. [...] We want to do something that will address the population's needs. Becoming involved in actions to recover heritage, to provide food, clothing, transport... (Manager 8, Construction, Wholesales and Distribution-Company 8)*

The main intention of our projects is social. We call them "social projects" because, for us, our reward lies in the hope that Timorese society will develop in educational terms. (Manager 5, Telecommunications service provider-Company 5)

The managers interviewed emphasize that the involvement of the corporate sector in CSR initiatives is particularly relevant in a context like that of East Timor. One should note that this territory has been independent for only 16 years and is dealing with persistent and complex economic difficulties (reduced rates of growth and economic sustainability, low productivity, and weak infrastructures) as well as social problems (poor human development, poverty, and unemployment), and for which the Government has insufficient response capacity [143–145].

As a result of existing CSR initiatives, the participants in this study thus specify various benefits to society, and especially focus on initiatives to promote better access to health (e.g., a mobile clinic for remote villages), as well as more and better education (e.g., awarding study grants) and access to technology. This last objective was highlighted by Manager 5:

For example, we have a large project called "Internet Community Center." Its mission is to contribute to the computerization of Timorese society so that it does, in fact, enter the era of computerization and technology. Through this CSR action, people can use the Internet; they can know what is happening in the world. (Manager 5, Telecommunications service provider-Company 5)

The group of participating managers acknowledges, however, that these affiliates experience restrictions in available resources, including those of time as well as the logistic problems associated to the country. These restrictions impair the response that could be provided by these affiliates to address all the requests presented by the stakeholder. This issue was expressed in the following comment:

The obstacles [in the implementation of CSR measures] are the same as those which companies are confronted with during their regular activities. There are no means, there is a lack of resources, and the planning of things has to be done several months ahead of time [to be successful]. Because some activities very often involve making purchases outside Timor and goods take 3 months to get here. There is often a sense of powerlessness because we could help, we know how to help, but we don't have the necessary equipment or the materials required to do so. (Manager 8, Construction, Wholesales and Distribution-Company 8)

These concerns of contributing to the well-being of the host society, which supplies important resources (namely a workforce and natural resources), suggest that there is a moral perspective in place. Besides these, practically all of the managers then chose to refer to economic objectives. In other words, the respective companies expect a return, generally in the long term, thus following a logic of reciprocity. There is, therefore, evidence of the instrumental perspective mentioned in the Stakeholder Theory [89]. Regarding this aspect, one must emphasize that some of these foreign branches are involved in actions which aim to "improve the population's physical well-being so that it is healthy and we can achieve good production" (Manager 4, Agrobusiness-Company 4). Another manager emphasized that the CSR actions undertaken to target environmental protection produced a positive impact on the quality of the main raw materials used by the company, the seeds of the coffee tree, which thus impacted on the quality of the coffee produced. Furthermore, the initiatives launched to train Timorese youngsters are considered to be a good investment; the purpose here is to contribute to a more qualified workforce to future integration in the company.

Several additional benefits for the companies are seen as ensuing from CSR initiatives which target employees. With regard to this aspect, one should highlight the measures taken to prevent risks at work and to minimize anti-ethical behaviors which could potentially jeopardize the company's productivity and image. The managers interviewed also considered that CSR activities benefit the company through the transfer of specific know-how [146, 147]. This represents an opportunity for employees to learn and to improve the staff's capacity which, in turn, contributes to good organizational performance [148]. As advocated by literature [41, 44, 149–151], various managers are of the opinion that people prefer to work for, and feel happier at, socially responsible companies; these are considered to offer organizational benefits related to a reduction in turnover, as well as greater loyalty and commitment on the part of employees, thus improving the company's capacity to attract talented workers. The following comment pointed to this perception:

The employees are really proud to work for a company which is so frequently involved with the community [through CSR measures]. It is obvious that, by being happier, these workers also do their jobs better because they feel well and they feel proud. (Manager 8, Construction, Wholesales and Distribution-Company 8)

The participating managers generally underlined the importance of the alignment of CSR strategies with those of the company's business. These were believed to conciliate and promote both social well-being and the company's economic objectives, as advocated by literature [37, 152, 153]. The discourse of some of the interviewees highlighted this concern:

Obviously, we cannot deviate from our general strategy. We should set up social responsibility activities or programs based on the plan of action determined by our general strategy. (Manager 4, Agrobusiness-Company 4)

This company is a Timorese company. Its existence in East Timor is also a service for the Timorese. So it cannot dodge the issue of the people's well-being, isn't it? [. . .] But, at the same time, the company also needs to make money through its commercial activities; it also has to invest in that. (Manager 5, Telecommunications service provider-Company 5)

As an example of this orientation which seeks to balance the various stakeholders' interests [98, 154] (namely, the shareholders' economic objectives and the host country's social expectations), the company managers from sectors such as those of energy, agriculture, and livestock production pointed to the importance of attenuating the risks associated to activities through concrete CSR measures. Their objective is, above all, to minimize the health and safety risks workers may be exposed to (by providing training in safety) as well as reduce potential impact on the environment.

3.4.3 Pursuing Legitimacy and the License to Operate

According to Institutional Theory, if an organization wishes to survive and prosper within a specific institutional framework, it must adjust its policies and practices to the social and cultural expectations of the particular context [100, 104]. The managers who participated in this research study explicitly acknowledged the need for the respective company to implement CSR measures which would address the demands of various local organizations and thus promote a good corporate image and reputation. These benefits are, in turn, understood to be promoters of both organizational legitimacy and the license to operate in East Timor as the host country. This aspect was also observed in previous studies undertaken in diverse national contexts [155–158]. The comments expressed by the following managers point to the perception of a link between the response to local needs and expectations, through CSR measures, and the acquisition of organizational legitimacy:

> *I believe and defend that the main benefit we obtain through our CSR initiatives has to do with our reputation. This has become a commitment for managers: gaining the positive perception of the community. We must ensure our reputation because, given that the organization is a bank, our business is especially about trust. (Manager 3, Banking-Company 3)*

> *Obviously we want to work within our community, improve our community, have a good working relationship with the community in which we find ourselves, and then build a harmonious relationship. (Manager 7, Agrobusiness-Company 7)*

> *Our behavior in CSR matters is extremely important because we want this company to be known as a reliable company, as a prestigious company. (Manager 8, Construction, Wholesales and Distribution-Company 8)*

> *No matter where we are, we always try to create activities that will support the community through our local partners, NGOs... By doing so, we contribute to the company's legitimacy. (Manager 6, Petroleum and Energy Distribution-Company 6)*

Of all the elements in the Timorese institutional environment which significantly influence the CSR decisions made by these foreign branches, the managers interviewed highlighted the role played by the Government. They explained that legislation in East Timor forces companies to redistribute a portion of its wealth to the community by means of CSR measures: "Based on the local law, our social responsibility actions can be summarized as being the profits that the company must withdraw and return to the local community". (Manager 2, Telecommunications service provider-Company 2). This result is similar to that of the study by Abaeian et al. [159] undertaken in another country in Asia, Malaysia.

Besides this perception that the Government is a promoter of "coercive isomorphism" [108] in CSR issues, it also emerges as a generator of normative isomorphism. As was explained by the managers interviewed, the Government's shortcomings and inability to respond seem to stimulate managers' ethical conscience and compel organizations to act in accordance with their sense of moral duty. Manager 6 emphasized this aspect:

> *We try to direct our social investment precisely toward the areas which the Government has identified as being the most deprived, and where we feel we can make a greater contribution. (Manager 6, Petroleum and Energy Distribution-Company 6)*

However, some managers explained that, despite the close relations and the partnership established with the Government, there are still persistent limitations and ambiguities in its regulations, which are seen as the result of the recent independence of the country. These culminate in barriers for CSR, which were also detected in previous studies undertaken in other developing countries [158, 160–162]. The following statements illustrated this perception:

> *Our relation with governmental entities is good but when we need to annually renew some certificates, for example, things are very difficult. [...] Some documents are rather unclear because they have two or three versions, for example, in English and Tetum, and the meaning differs. Therefore, we occasionally experience great difficulties with the laws and bureaucracy involved. (Manager 1, Logistics and Shipping-Company 1)*

> *We want to help the government implement e-government. It's to save money, and not use so much paper. If we implement e-government, we will be more efficient and simultaneously contribute to the protection of the environment. (Manager 5, Telecommunications service provider-Company 5)*

On the other hand, it is the pressure exerted by other Timorese institutions, namely, those which are associated to education and religion, which seems to constitute another mechanism for the promotion of "coercive isomorphism" in CSR issues. This is indicated by Manager 8 (Construction, Wholesales, and Distribution-Company 8): "The community is very active, teachers, nuns... and they are not ashamed of asking us for help, which is very good."

3.4.4 Adjusting Parent-company Policies to Local Needs

According to the managers interviewed, the involvement in CSR measures of companies in East Timor is significantly stimulated by their respective headquarters abroad, namely, in the definition of a general strategy. This was also presented in the study by Ruud [163] with regard to the context of India. The influence exercised by the respective headquarters is especially noteworthy. Due to the fact that the parent company of most of these affiliates is located in countries where social and cultural expectations determine CSR policies, these are more expressive than those practiced by local Timorese companies. In addition to the roles played by Timorese legislation and some institutions in the country, foreign headquarters have thus proved to be fundamental mechanisms of "coercive isomorphism" [104] when dealing with CSR issues in this host country.

The managers interviewed guaranteed, however, that the specific nature of the Timorese context is considered by the Boards of Administration and by the local shareholders in order to ensure policies and practices which are more suited to the local culture and the community's needs: "The culture in East Timor is different, different things confront us, as well as specific needs (Manager 2, Telecommunications service provider-Company 2).

In real terms, while the definition of the CSR strategy is decided by the respective foreign headquarters, the managers interviewed confirmed that the parent company's actions are clearly less significant when translated into concrete CSR policies and practices, as well as their respective implementation in Timorese branches. This is especially true when organizations locally define which stakeholders should be prioritized, and decide upon the respective measures and practices that will ensure a better response to local needs. These results – similar to those obtained in the studies undertaken by Jamali [32] and Cruz and Boehe [164], in Lebanon and Brazil, respectively – reflect what Yin and Zhan [44] refer to as distinct dimensions of pressure in the development of CSR measures: (i) the macro dimension, which relates to pressures exerted by the global regulating entity, or foreign headquarters; and (ii) the micro dimension, which refers to local social, cultural, and legislative pressures.

The following comments expressed by some of the managers interviewed illustrate this exercise of a local adaptation of global guidelines in the case of CSR matters:

> *The shareholders [local] know the country, they know what its needs are…Therefore, all those who work in East Timor, the*

shareholders themselves, motivate the company to relate to society in a positive way. (Manager 8, Construction, Wholesales and Distribution-Company 8)

The decision is made at headquarters, but its implementation and action are carried out in a local context. [...] Our targets, as well as assessments, are defined by headquarters, but when it comes to their implementation, action and evaluation, we do it here and then report to headquarters. However, every 2 years, they come here to do an audit. So, there is great control, but we are free to make decisions. There is, therefore, some flexibility. (Manager 1, Logistics and Shipping-Company 1)

Headquarters keeps track, it's like having a technical board. But they don't stay here permanently, and they don't get involved in decision-making. What they do is give us advice, but we then implement it. So we're not highly dependent on them. They always listen to us. (Manager 4, Agrobusiness-Company 4)

Various managers concluded by explaining that Timorese affiliates are especially free to make decisions regarding occasional and low-budget CSR projects; however, in the case of larger projects, which take place regularly and involve large budgets, headquarters' decisive participation is required. One should point out that this model of combining global and local decisions is acknowledged to be effective in the promotion of legitimacy and the license to operate in host countries, as well as in the important access to resources [30].

3.5 Conclusion

Literature has emphasized the constant lack of consensus in establishing a general definition of CSR [8, 9]; it also points to the unique nature of this concept, as well as to its application and importance in specific national contexts [18, 19]. Although several research studies have focused on diverse countries in Asia [39], which relate CSR practices with local social, cultural and political characteristics, less attention has been directed to the study of this subject in East Timor. The purpose of the present study was to bridge this gap by presenting the perceptions of 10 branch managers working for Foreign MNCs with responsible positions in CSR. This national context is particularly unique: the country in question is still rather young and has been confronted with complex social and economic problems, which the government has not been able to address completely. These issues have heightened the pertinence of CSR initiatives by companies with more resources and knowledge in this domain, namely, the affiliates of MNCs based in countries where this area of activity is more developed and involves greater social expectations.

Focusing on the Timorese context, this study is anchored on the results of in-depth interviews with managers in positions which include responsibilities in CSR matters. It points to the existence of specific social, economic, and cultural reasons in the definition of stakeholders and their relative importance. The internal stakeholder group mentioned consists of shareholders and employees. However, while literature usually acknowledges employees as key internal stakeholders [82, 141], their weak bargaining power in the Timorese context seems to discourage more expressive CSR measures directed at this group. On the other hand, the list of external stakeholders includes the Government and public institutions, NGOs, educational and religious institutions, consumers/customers, business partners, as well as the general community. The relevant role played by general society in a company's survival/prosperity is acknowledged, namely, in order to obtain resources; society's needs thus exert a significant impact on the definition of CSR measures. These seek, above all, to contribute to the attenuation of persistent and complex social problems which the Timorese government cannot address adequately, namely, in the areas of health, education, and access to technology. Environmental concerns are less obvious and emerge mainly when companies acknowledge the negative environmental impact of their activity.

The Government is assessed as being a "priority external" or "dominant" stakeholder due to its accumulated roles in East Timor. Besides being a regulator, the Government is also a shareholder and the companies' main customer. This stakeholder has thus proved to be an important player in the implementation of the coercive isomorphism of CSR issues in this country. At the same time, the Government seems to tend toward a normative type of isomorphism. Due to the fact that it is unable to address certain social and economic domains nor respond to the needs presented by educational and religious institutions, the Government has fostered managers' ethical conscience, driving companies to act in the most deprived areas of society in accordance with their moral duty.

These reasons of ethical or moral consideration are linked to others of a more strategic nature; this emerges in the justifications provided by managers to explain the growing involvement of the respective affiliates in CSR initiatives, where the existence of the instrumental perspective of the Stakeholder Theory then becomes obvious. The alignment of a CSR strategy with that of the company's business is considered essential to conciliate and promote both the objectives of social well-being and economic issues. The positive impact of developing CSR measures to address the community's well-being and to ensure the promotion of the companies' good image and reputation is also firmly acknowledged. These results are considered to be fundamental to the development and maintenance of organizational legitimacy and the license to operate in East Timor.

The involvement of the affiliates of foreign-based companies in CSR actions in East Timor appears to be significantly stimulated by their respective headquarters abroad, namely, in relation to the definition of the respective general strategy, so that this constitutes another important mechanism of coercive isomorphism regarding this issue. The specific characteristics of the Timorese context are, however, considered by the Administration Boards and by local shareholders, who thus guarantee that the CSR strategy is converted into policies and practices suited to the local culture and needs. One should point out that this model, which combines global and local decisions, is acknowledged to be effective in promoting legitimacy and maintaining a license to operate in host countries [30].

However, there are still some hurdles in the development and implementation of CSR measures in this young Asiatic country, which are associated to the specific features of the Timorese context. These reside chiefly in the ambiguities of Government regulations and issues of a bureaucratic nature, as well as in problems of logistics which complicate access to the resources which are essential to the response given to stakeholders' requests.

Future suggested research should focus on the Timorese context, directing special attention to tangible CSR practices implemented by the affiliates of foreign-based MNCs. The purpose would be to gain a better understanding of the way in which these balance out the directives issued by headquarters with the fundamental adjustment of these to local characteristics, namely, with regard to stakeholders' requests and specific needs.

References

[1] Bowen, H. (1953). *Social Besponsibilities of the Businessman*, New York: NY, Harper and Row.
[2] Carroll, A. B. (1999). Corporate social responsibility: evolution of definitional construct. *Bus. Soc.* 38, 268–295.

[3] Friedman, M. (1970). The social responsibility of business is to increase its profits. *N. Y. Times Mag.* 13, 32–33.

[4] Keinert, C. (2008). *Corporate social responsibility as an international strategy*, Leipzig: Germany, Phsica-Verlag A Springer Company.

[5] Maon, F., Lindgreen, A., and Swaen, V. (2010). Organizational stages and cultural phases: a critical review and a consolidative model of corporate social responsibility development. *Int. J. Manag. Rev.* 12, 20–38.

[6] Garriga, E., and Melè, D. (2004). Corporate responsibility theories: mapping the territory. *J. Bus. Ethics* 53, 51–71.

[7] Jamali, D. A. (2008). Stakeholder approach to corporate social responsibility: a fresh perspective into theory and practice. *J. Bus. Ethics* 82, 213–231.

[8] Schwartz, M., and Carroll, A. (2007). Integrating and unifying competing frameworks: the search for a common core in the business and society field. *Bus. Soc.* 47, 148–186.

[9] Carroll, A. B., and Shabana, K. (2010). The business case for corporate social responsibility: a review of concepts, research and practice. *Int. J. Manag. Rev.* 12, 85–105.

[10] Idemudia, U. (2011). Corporate social responsibility and developing countries: moving the critical CSR research agenda in Africa forward. *Prog. Dev. Stud.* 11, 1–18.

[11] Scherer, A., Palazzo, G., and Matten, D. (2009). Introduction to the special issue: globalization as a challenge for business responsibilities. *Bus. Ethics Q.* 19, 327–347.

[12] Prieto-Carron, M., Lund-Thomsen, P., Chan, A., Muro, A., and Bhushan, C. (2006). Critical perspective on CSR and development: what we know, what we don't know, and what we need to know. *Int. Aff.* 82, 977–987.

[13] Blowfield, M., and Frynas, J. G. (2005). Setting new agenda: critical perspectives on corporate social responsibility in the developing world. *Int. Aff.* 81, 499–513.

[14] Wong, A., Long, F., and Elankumaran, S. (2010). Business students' perception of corporate social responsibility: the United States, China, and Italia. *Corp. Soc. Responsib. Env. Manag.* 17, 299–310.

[15] Campbell, J. L. (2007). Why would corporations behave in socially responsible ways? An institutional theory of corporate social responsibility. *Acad. Manag. Rev.* 32, 946–967.

[16] Hilman, A., and Wan, W. (2005). The determinants of MNE subsidiaries political strategies: evidence of institutional duality. *J. Int. Bus. Stud.* 36, 322–340.

[17] Maignan, I. (2002). Consumers' perceptions of corporate social responsibilities: a cross-cultural comparison. *J. Bus. Ethics* 20, 57–72.

[18] Matten, D., and Moon, J. (2008). ≪Implicit≫ and ≪explicit≫ CSR: a conceptual framework for a comparative understanding of corporate social responsibility. *Acad. Manag. Rev.* 33, 404–424.

[19] Baughn, C. C., Bodie, N. L., and McIntosh, J. C. (2007). Corporate social and environmental responsibility in Asian countries and other geographical regions. *Corp. Soc. Responsib. Env. Manag.* 14, 189–205.

[20] Kusku, F., and Zarkada-Fraser, A. (2004). An empirical investigation of corporate citizenship in Australia and Turkey. *Br. Acad. Manag.* 15, 57–72.

[21] Welford, R. (2005). Corporate social responsibility in Europe, North America and Asia: 2004 survey result. *J. Corp. Citizensh.* 17, 33–52.

[22] Chambers, E., Chapple, W., Moon, J., and Sullivan, M. (2003). CSR in Asia: a seven country study of CSR website reporting. *Paper Presented at the International Centre for Corporate Social Responsibility*, Nottingham.

[23] Maignan, I., and Ralston, D. A. (2002). Corporate social responsibility in Europe and the U.S.: insights from businesses self-presentations. *J. Int. Bus. Stud.* 33, 497–514.

[24] Bagire, V., Tusiime, I., Nalweyiso, G., and Kokooza, J. (2011). Contextual environment and stakeholder perception of corporate social responsibility practice in Uganda. *Corp. Soc. Responsib. Env. Manag.* 18, 102–109.

[25] Halme, M., Roome, N., and Dobers, P. (2009). Corporate responsibility: reflections on contexts and consequences. *Scand. J. Manag.* 25, 1–9.

[26] Sharma, B. (2013). *Contextualizing CSR in Asia: corporate social responsibility in Asian economics*. Singapore: Lien Centre for Social Innovation.

[27] Chapple, W., and Moon, J. (2005). Corporate social responsibility (CSR) in Asia: a seven-country study of CSR web site reporting. *Bus. Soc.* 44, 415–441.

[28] Lindgreen, A., Swaen, V., and Harness, D. (2011). The role of high potentials in integrating and implementing corporate social responsibility. *J. Bus. Ethics* 99, 73–91.

[29] Yang, S., and Rivers, C. (2009). Antecedents of CSR practices in MNCs' subsidiaries: a stakeholder and institutional perspective. *J. Bus. Ethics* 86, 155–169.

[30] Muller, A. (2006). Global versus local CSR strategies. *Eur. Manag. J.* 24, 189–198.

[31] Jamali, D. (2010). The CSR of MNC subsidiaries in developing countries: global, local, substantive or diluted? *J. Bus. Ethics* 93, 181–200.

[32] Zhu, Q., and Zhan, Q. (2015). Evaluating practices and drivers of corporate social responsibility: the China context. *J. Clean. Prod.* 100, 315–324.

[33] Pastrana, N. A., and Sriramesh, K. (2014). Corporate social responsibility: perceptions and practices among SMEs in Colombia. *Public Relat. Rev.* 40, 14–24.

[34] Belal, A. R., and Cooper, S. (2011). The absence of corporate social responsibility reporting in Bangladesh. *Critical Perfect. Account.* 22, 654–667.

[35] Muthuri, J. N., and Gilbert, V. (2011). An institutional analysis of corporate social responsibility in Kenya. *J. Bus. Ethics* 98, 467–483.

[36] Muwazir, M. R. (2011). *Corporate social responsibility in the context of financial services sector in Malaysia.* Ph D. thesis, dissertation, Cardiff University, Department of Account and Finance United Kingdom.

[37] Jamali, D., and Mirshak, R. (2007). Corporate social responsibility (CSR): theory and practice in a developing country context. *J. Bus. Ethics* 72, 243–262.

[38] Luken, R. A. (2006). Where is developing country industry in sustainable development planning? *Sustain. Dev.* 14, 46–61.

[39] Srinivasan, V. (2011). Business ethics in the South of East Asia. *J. Bus. Ethics* 104, 73–81.

[40] Belal, A. R. (2001). A study of corporate social disclosure in Bangladesh. *Manag. Audit. J.* 16, 274–289.

[41] Frynas, J. G. (2005). The false development promise of corporate social responsibility: evidence from multinational oil companies. *Int. Aff.* 81, 581–598.

[42] Kimber, D. and Lipton, P. (2005). Corporate Governance and Business Ethics in the Asia-Pacific region. *Business & Society*, 44: 178–210.

[43] Welford, R. and Frost, S. (2006). Corporate social responsibility in Asian supply chains. *Corp. Soc. Responsib. Env. Manag.* 13, 166–176.

[44] Yin, J., and Zhang, Y. (2012). Institutional dynamics and corporate social responsibility (CSR) in an emergency country context: evidence from China. *J. Bus. Ethics* 111, 301–316.

[45] Gao, Y. (2011). Philanthropic disaster relief giving as a response to institutional pressure: evidence from China. *J. Bus. Res.* 64, 1377–1382.

[46] Tang, L. and Li, H. (2009). Corporate social responsibility communication of Chinese and global corporations in China. *Public Relat. Rev.* 35, 199–212.

[47] Wang, J., and Chaudhri, V. (2009). Corporate social responsibility engagement and communication by Chinese companies. *Public Relat. Rev.* 35, 247–250.

[48] Kulkarni, M. S. (2015). A study of the CSR policies and practices of Indian companies. *J. Contemp. Res. Manag.* 17:26.

[49] Rani, P., and Khan, M. (2015). Changing scenario of corporate social responsibility (CSR) in an era of globalization. *Int. J. Appl. Res.* 1, 111–114.

[50] Govindan, K., Kannan, D., and Shankar K. M. (2014). Evaluating the drivers of corporate social responsibility in the mining industry with multi-criteria approach: a multi-stakeholder perspective. *J. Clean. Prod.* 84, 214–232.

[51] Arevalo, J. A., and Aravind, D. (2011). Corporate social responsibility practices in India: approach, drivers, and barriers. *Corp. Gov.* 11, 399–414.

[52] Pradhan, S., and Ranjan, A. (2010). Corporate social responsibility in rural development: evidences from India. *Sch. Dr. Stud. J.* 139:147.

[53] Ismail, M., Johar, R. F. A., Rasdi, R. M., and Alias, S. N. (2014). School as stakeholder of corporate social responsibility program: teacher's perspective on outcome in school development. *Asia Pac. Edu. Responsib.* 23, 321–331.

[54] Muwazir, M. R., and Hadi, N. A. (2014). "Motivation for CSR Practices: Evidence From Financial Services Industry," in *Proceedings of the 2nd IBEA – International Conference on Business Economics and Accounting*, Hongkong, 26–28.

[55] Yam, S. (2013). The practice of corporate social responsibility by Malaysian developers. *Prop. Manag.* 31, 76–91.

[56] Zabin, I. (2013). An investigation of practicing Carroll's pyramid of corporate social responsibility in developing countries: an example of Bangladesh ready-made garments. *IOSRJ. Bus. Manag.* 12, 75–81.

[57] Musdiana, Salleh., M., Nasbiah, and Wahid, A. (2012). Corporate social responsibility (CSR) in Malaysian banking industry: an analysis through website of six banking institutions. *Elixir Mark. Manag.* 50, 10225–10234.

[58] Hossain, M. M., Rowe, A. L., and Quaddus, M. A. (2012). "Drivers and barriers of corporate social and environmental reporting (CSER) practices in a developing country: evidence from bangladesh," in *Proceedings of the 10th Interdisciplinary Perspectives on Accounting (IPA) Conference*, Cardiff, UK.

[59] Hossain, M., and Rowe, L. A. (2011). "Enablers for corporate social and environmental responsibility (CSER) practices: Evidence from Bangladesh," in *Proceedings of the 10th CSEAR Australasian Conference (Conference on Social and Environmental Accounting Research)*. University of Tasmania, Launceston.

[60] Quazi, A., Rahman, Z., and Keating, B. (2007). "A developing county perspective of corporate social responsibility: A test case of Bangladesh," in *Proceeding of the Australian and New Zealand Marketing Academy Conference*. Duneding, New Zealand: New Zealand Marketing Academy.

[61] Munasinghe, M. A. T. K., and Malkumari, A. P. (2012). Corporate social responsibility in small and medium enterprises (SME) in Sri Lanka. *J. Emerg. Trends Edu. Res. Policy Stud.* 3, 168–172.

[62] Kalyar, M. N., Rafi, N., and Kalyar, A. N. (2013). Factors affecting corporate social responsibility: an empirical study. *Sys. Res. Behav. Sci.* 30, 495–505.

[63] Trang, H. N. T., and Yekini, L. S. (2014). Investigating the link between CSR and financial performance: evidence from Vietnamese listed companies. *Br. J. Arts Soc. Sci.* 17, 85–101.

[64] Fauzi, H. (2014). The Indonesian executives perspective of CSR practices. *Issues Soc. Env. Account.* 8, 171–181.

[65] Asniwaty, B. (2010). Evaluasi pelaksanaan corporate social responsibility (CSR) PT Pupuk Kaltim. *Jurnal Eksis* 6, 1267–1273.

[66] Rani, P., and Khan, M. (2015). Changing scenario of corporate social responsibility (CSR) in an era of globalization. *Int. J. Appl. Res.* 1, 111–114.

[67] Abdul, M. Z., and Ibrahim, S. (2002). Executive and management attitudes towards corporate social responsibility in Malaysia. *Corp. Gov.* 2, 10–16.

[68] Chen, F. Y., Chang, Y. H. and Lin, Y. H. (2012). Customer perceptions of airline social responsibility and its effect on loyalty. *J. Air Trans. Manag.* 20, 49–51.

[69] Maon, F., Lindgreen, A., and Swaen, V. (2009). Designing and implementing corporate social responsibility: an integrative framework grounded in theory and practice. *J. Bus. Ethics* 87, 71–89.

[70] Aguinis, H., and Glavas, A. (2012). What we know and don't know about corporate social responsibility: a review and research agenda. *J. Manag.* 38, 932–968.

[71] Husted, B. W., and Allen, D. B. (2006). Corporate social responsibility in the multinational enterprise: strategic and institutional approaches. *J. Int. Bus. Stud.* 37, 838–849.

[72] Basu, K., and Palazzo, G. (2008). Corporate social responsibility: a process model of sensemaking. *Acad. Manag. Rev.* 33, 122–136.

[73] Carroll, A. B. (1993). *Business and Society: Ethics and Stakeholder Management.* 2nd Edn. Cincinnati, OH: South-Western Publishing.

[74] McWilliams, A., and Siegel, D. (2001). Corporate social responsibility: A theory of the firm perspective. *Acad. Manag. Rev.* 26, 117–127.

[75] Freeman, R. E. (1984). *Strategic Management: A Stakeholder Approach.* Boston: Pitmann.

[76] Freeman, R. E. (2002). "Stakeholder theory of the modern corporation," in *Ethical Issues in Business: A Philosophical Approach*, 7th Edn, eds T. Donaldson. P. Werhane P (Englewood Cliffs, NJ: Prentice Hall), 38–48.

[77] Pirsch, J., Gupta, S., and Grau, S. L. (2007). A framework for understanding corporate social responsibility programs as a continuum: an exploratory study. *J. Bus. Ethics* 70, 125–140.

[78] Post, J. E., Lawrence, A. T., and Weber, J. (2002). *Business and Society: Corporate Strategy Public Policy Ethics.* Boston, MA: McGraw-Hill.

[79] Antolìn, M. N. (2008). La difusión de las prácticas de responsabilidad social en las empresas multinacionales. *Pecvinia*, 1, 33–64.

[80] Kakabadase, N. K., Rozuel, C., and Lee-Davis, L. (2005). Corporate social responsibility and stakeholder approach: a conceptual review. *Int. J. Bus. Gov. Ethics* 1, 277–302.

[81] Freeman, R. E., and David, L. R. (1983). Stockholder and stakeholders: a new perspective on corporate governance. *Calif. Manag. Rev.* 25, 88–106.

[82] Clarkson, M. (1995). A stakeholder framework for analyzing and evaluating corporate social responsibility. *Acad. Manag. Rev.* 20, 92–118.

[83] Öberseder, M., Schlegelmilch, B. B., and Murphy, P. E. (2013). CSR practices and consumer perceptions. *J. Bus. Res.* 66, 1839–1851.

[84] Carroll, A. B. (1991). The pyramid of corporate social responsibility: toward the moral management of organizations stakeholder. *Bus. Horiz.* 34, 39–48.

[85] Zhao, Z.-Y., Zhao, X.-J., Davidson, K., and Zuo, J. (2012). A corporate social responsibility indicator system for construction enterprises. *J. Clean. Prod.* 29–30, 277–289.

[86] Waddock, S. and Smith, N. (2000). Relationships: The real challenge of corporate global citizenship. *Bus. Soc. Rev.* 105, 47–62.

[87] Jones, G. R. (2010). *Teoria Das Organizações.* eds. L. Luciane and D. Viera. São Paulo: Pearson.

[88] Freeman, R. E., and Liedtkn, J. (1997). Stakeholder capitalism and the value chain. *Eur. Manag. J.* 15, 286–296.

[89] Donaldson, T., and Preston, L. E. (1995). The stakeholder theory of the corporation: concepts, evidence and implications. *Acad. Manag. Rev.* 20, 65–91.

[90] Frooman, J. (1999). Stakeholder influence strategies. *Acad. Manag. Rev.* 24, 191–205.

[91] Phillips, R. (2003). Stakeholder legitimacy. *Bus. Ethics Q.* 13, 25–41.

[92] Belal, A. R., and Owen, D. (2007). The views of corporate managers on the current state of, and future prospects for social reporting in Bangladesh: An engagement based study. *Account. Audit. Account. J.* 20, 472–494.

[93] Deegan, C. (2009). *Financial Accounting Theory.* Sydney, NSW: McGraw-Hill.

[94] Tsoi, J. (2010). Stakeholders' perception and future scenarios to improve corporate social responsibility in Hong-Kong and mainland China. *J. Bus. Ethics* 91, 391–404.

[95] Deegan, C. and Jeffry, U. (2006). *Financial Accounting Theory.* Berkshire: McGraw-Hill Education.

[96] Roberts, R. W. (1992). Determinants of corporate social responsibility disclosure: an application of stakeholder theory. *Account. Organ. Soc.* 17, 595–612.

[97] O'Riordan, L., and Fairbrass, J. (2014). Managing CSR stakeholder engagement: a new conceptual framework. *J. Bus. Ethics* 125, 121–145, 2014.

[98] Dufrene, U., and Wong, A. (1996). Stakeholder versus stockholder and financial ethics: Ethics to whom? *Manag. Finance* 22, 1–11.

[99] Mitchell, R. K., Agle, B. R., and Wood, D. J. (1997). Toward a theory of stakeholder identification and salience: defining the principle of who and what really counts. *Acad. Manag. Rev.* 22, 853–886.

[100] Zucker, L. G. (1977). The role of institutionalization in cultural persistence. *Am. Sociol. Rev.* 42, 726–743.

[101] Sánchez-Fernandes, M. D. (2014). Institutional context of hotel social responsibility in the Euro-region: a factorial analysis. *Rev. Ocio Turismo* 7, 106–119.

[102] Scott, W. R. (1995). *Institutions and Organizations.* Thousand Oaks, CA: Sage.

[103] Lee, M. D. (2011). Configuration of external influences: the combined effects of institutions and stakeholders on corporate social responsibility strategies. *J. Bus. Ethics* 102, 281–298.

[104] Suchman, M. C. (1995). Managing legitimacy: strategic and institutional approaches. *Acad. Manag. Rev.* 20, 571–610.

[105] DiMaggio, P. J., and Powell, W. W. (1983). The iron cage revisited: institutional isomorphism and collective rationality in organizational fields. *Am. Sociol. Rev.* 48, 147–160.

[106] Scott, W. R. (2008). Approaching adulthood: the maturing of institutional theory. *Theory Sociol.* 37, 427–442.

[107] Mahmood, M., and Humphrey, J. (2013). Stakeholder expectation of corporate social responsibility practices: a study on local and multinational corporations in Kazakhstan. *Corp. Soc. Respons. Environ. Manag.* 20, 168–181.

[108] Fifka, M. S., and Pobizhan, M. (2014). An institutional approach to corporate social responsibility in Russia. *J. Clear Prod.* 82, 192–201.

[109] Galaskiewicz, J. (1991). "Making corporate actors accountable: Institution-building," in *The New Institutionalism in Organizational*

Analysis, eds W. W. Powell and P. J. DiMaggio (Chicago: University of Chicago Press), 293–310.

[110] Aguilera, R. V., Rupp, D. E., Williams, C. A., and Ganapathi, J. (2007). Putting the S back in corporate social responsibility: a multilevel theory of social change in organizations. *Acad. Manag. Rev.* 32, 836–863.

[111] Tan, J., and Wang, L. (2011). MNC strategic responses to ethical pressure: an institutional logic perspective. *J. Bus. Ethics* 98, 373–390.

[112] O'Donovan, G. (2000). *Legitimacy Theory as an Explanation for Environmental Disclosures*, Ph.D. dissertation, Victoria University of Technology, Melbourne.

[113] Scott, W. R. (2001). *Institutions and Organizations.* 2nd Edn. Thousand Oaks, CA: Sage.

[114] Zimmerman, M. A., and Zeitz, G. J. (2002). Beyond survival: achieving new venture growth by building legitimacy. *Acad. Manag. Rev.* 27, 414–431.

[115] Dìez Martìn, F., Blanco González, A., and Prado Román, C. (2010). Medición de la legitimidad organizativa: El caso de las Sociedades de Garantìa Recìproca. *Cuad. Econ. Direc. Empresa* 12, 115–143.

[116] Handelman, J. M., and Arnold, S. J. (1999). The role of marketing actions with a social dimension: appeals to the institutional environment. *J. Mark.* 63, 33–48.

[117] O'Donovan, G. (2002). Environmental disclosure in the annual report: extending the applicability and predictive power of legitimacy theory. *Account. Audit. Account. J.* 15, 344–371.

[118] Islam, M. A., and Deegan, C. (2008). Motivations for an organization within a developing country to report social responsibility information: evidence from Bangladesh. *Account. Audit. Account. J.* 21, 850–874.

[119] Bansal, P., and Clelland, I. (2004). Talking trash: legitimacy, impression management, and unsystetamtic risk in the context of the natural environment. *Acad. Manag. J.* 47, 93–103.

[120] Claasen, C., and Roloff, J. (2012). The link between responsibility and legitimacy: the case of De Beers in Namibia. *J. Bus. Ethics*, 107, 379–398.

[121] Du, S., and Vieira, E. T. (2012). Striving for legitimacy through corporate social responsibility: insights from oil companies. *J. Bus. Ethics* 110, 413–427.

[122] Deegan, C. (2002). Introduction: the legitimizing effect of social and environmental disclosure: a theoretical foundation. *Account. Audit. Account. J.* 15, 282–311.

[123] Deegan, C., and Rankin, M. (1996). Do Australian companies report environmental news objectively?: an analysis of environmental disclosures by firms prosecuted successfully by the environmental protection authority. *Account. Audit. Account. J.* 9, 50–67.

[124] Mobus, J. L. (2005). Mandatory environmental disclosure in a legitimacy theory context. *Account. Audit. Account. J.* 18, 492–517.

[125] An, Y., Davey, H., and Eggleton, I. R. C. (2011). Towards a comprehensive theoretical framework for voluntary IC disclosure. *J. Intel. Capital* 12, 571–585.

[126] Laan, S. L. V. D. (2004). *The Role of Theory in Explaining Motivation for Corporate Social Disclosures: Voluntary Disclosure vs Solicited Disclosures.* Paper Presented at Asia Pacific Interdisciplinary Research in Accounting Conference, Singapore.

[127] Belkoui, A., and Karpik, P. G. (1998). Determinants of the corporate decision to disclose social information. *Account. Audit. Account. J.* 1, 36–51.

[128] Woodward-Clyde (1999). Key opportunities and risks to new zealand's export trade from green market signals, final. *Paper, Sustainable Management Fund Project 6117. New Zealand Trade and Development Board*, Auckland.

[129] Neu, D. Warsame, H., and Pedwell, K. (1998). Managing public impressions: environmental disclosures in annual reports. *Account. Organ. Soc.* 23, 265–282.

[130] Tilt, C. A. (1998). *The content and disclosure of corporate environmental policies: An Australian study.* Paper presented at the First Asian Pacific Interdisciplinary Research in Accounting Conference. Sydney, NSW.

[131] Palazzo, G., and Richter, U. (2005). CSR business as usual? the case of the tobacco industry. *J. Bus. Ethics*, 61, 387–401.

[132] Wei, Z., Shen, H., Zhou, K. Z., and Li, J. J. (2017). How does environmental corporate social responsibility matter in a dysfunctional institutional environment? Evidence from China. *J. Bus. Ethics*, 140, 209–223.

[133] Amaeshi, K. M., Adi, B. C., Ogbechie, C., and Amao, O. O. (2006). Corporate social responsibility in Nigeria: Westeren mimicry or indigenous influences? *J. Corp. Citizen.* 24, 83–99.

[134] Black, J. A., and Champion, D. J. (2009). *Metode dan Masalah Penelitian Sosial.* Bandung: Rafica Aditama.

[135] Teddlie, C., and Tashakkori, A. (2009). *Foundations of Mixed Methods Research: Integrating Quantitative and Qualitative Approaches to the Social and Behavioral Sciences.* Thousand Oaks, CA: Sage.

[136] Creswell, J. W., and Plano Clark, V. (2011). *Designing and conducting mixed methods research.* 2nd Edn. Thousand Oaks, CA: Sage Publication.

[137] Kaur, M. (2016). Application of mixed method approach in public health research. *Indian J. Commun. Med.* 41, 93–93.

[138] Gond, J. P., Kang, N., and Moon, J. (2011). The government of self-regulation: on the comparative dynamics of corporate social responsibility. *Econ. Soc.* 40, 640–671.

[139] Yin, J., and Jamali, D. (2016). Strategic corporate social responsibility of multinational companies subsidiaries in emerging markets: evidence from China. *Long Range Plann.* 49, 541–558.

[140] Fernando, S., Lawrence, S., Kelly, M., and Arunachalam, M. (2015). RSC practices in Sri-Lanka: an exploratory analysis. *Soc. Respons. J.* 11, 868–892.

[141] Turker, D. (2009). How corporate social responsibility influences organizational commitment. *J. Bus. Ethics* 89, 189–204.

[142] Ministery of Economic and Development (MED). (2012). *Sustainable Development in Timor-Leste: National Report to The United Nations Conference on Sustainable Development (UNCSD) On the run up to Rio+20,* 2012. Maldives: Ministery of Economic and Development.

[143] Asian Development Bank (ADB). (2016). *Country Partnership Strategy: Timor-Leste.*

[144] Harmadi, S. H. B., and Gomes, R. A. (2013). Developing Timor-Leste's non-oil economy: challenges and prospects. *J. South. Asian Econ.* 30, 309–321.

[145] World Bank and Directorat of National Statistics (WB & DNS). (2008). *Timor-Leste: Poverty in a young nations. Preliminary Draft Report, Dili-Timor Leste, 2008.* Available: http://pascal.iseg.utl.pt/~cesa/TL-Poverty-in-a-young-nation-25-Nov-2008.pdf

[146] Ismail, M. (2009). Corporate social responsibility and its role in community development: An international perspective. *J. Int. Soc. Res.* 2, 199–209.

[147] Ite, U. E. (2004). Multinationals and corporate social responsibility in development countries: a case study of Nigeria. *Corp. Soc. Environ. Manag.* 11, 1–11.

[148] Heslin, P. A., and Ochoa, J. D. (2008). Understanding and developing strategic corporate social responsibility. *Organ. Dynam.* 32, 125–144.

[149] Zhu, Q., Hang, Y., Liu, J., and Lai, K. H. (2014). How is employee perception of organizational efforts in corporate social responsibility related to their satisfaction and loyalty towards developing harmonious society in Chinese enterprises? *Corp. Soc. Respons. Environ. Manag.* 21, 28–40.

[150] Bhattacharya, C. B., Sen, S., and Korschun, D. (2008). Using corporate social responsibility to win the war for talent. *Sloan Manag. Rev.* 49, 37–44.

[151] Bevan, S., Isles, N., Emery, P., and Hoskins (2004). *Achieving high performance: RSC at the heart of business.* London: The Work Foundation.

[152] Vogel, D. (2005). *The Market for Virtue: Potential and Limits of Corporate Social Responsibility.* Harrisonburg: The Brookings Institution.

[153] Porter, M. E., and Kramer, M. R. (2002). The competitive advantage of corporate philanthropy. *Harv. Bus. Rev.* 80, 56–68.

[154] O'Riordan, L. (2010). *Perspectives on corporate social responsibility (CSR): Corporate approaches to stakeholder engagement in the pharmaceutical industry in the UK and Germany.* Ph.D. thesis, Bradford University School of Management, Bradford.

[155] Allahverdiyeva, T. (2016). Corporate social responsibility: mutual expediency of transnational companies and developing countries. *Open J. Bus. Manag.* 4, 355–360.

[156] Askeroğlu, E. D., and Bahar, Z. (2014). Contribution of foundations to reputation in corporate social responsibility applications: vodafone Turkey Foundation review. *Int. J. Econ. Pract. Theor.* 4, 718–725.

[157] Coppa, M., and Sriramesh, K. (2013). Corporate social responsibility among SMEs in Italy. *Public Relat. Rev.* 39, 30–39.

[158] Visser, W. (2008). "Corporate social responsibility in developing countries," in *The Oxford Handbook of Corporate Social Responsibility*, eds A. Crane, A. McWilliams, D. Matten, J. Moon, and D. Siegel (Oxford: Oxford University Press), 473–499.

[159] Abaeian, V., Yeoh, K. K., and Khong, K. W. (2014). An exploration of RSC initiatives undertaken by Malaysian hotels: underlying motivations from a managerial perspective. *Proc. Soc. Behav. Sci.* 144, 423–432.

[160] Buhmann, K. (2005). Corporate social responsibility in China: current issues and their relevance for implementation of law. *Copenhag. J. Asian Stud.* 22, 62–91.

[161] Gupta, D. A. (2007). Social responsibility in India towards global compact approach. *Int. J. Soc. Econ.* 34, 637–663.

[162] Hilson, G. (2012). Corporate social responsibility in the extractive industries: experiences from developing countries. *Resourc. Policy* 37, 131–137.

[163] Ruud, A. (2002). Environmental management of transnational corporations in India: are TNCs creating islands of environmental excellence in a Sea of Dirt? *Bus. Strat. Environ.* 11, 103–119.

[164] Cruz, L. B., and Boehe, D. M. (2010). How do leading retail MNCs leverage CSR globally? Insights from Brazil. *J. Bus. Ethics* 91, 243–263.

4

Gender Diversity and Equality in the Boardroom: Impacts of Gender Quota Implementation in Portugal

Mara Sousa and Maria João Santos

School of Economics and Management, University of Lisbon, Lisboa, Portugal

Abstract

Given the importance of gender equality to international economics and sustainable economic and social development, this article addresses the gender imbalances in senior company board decision-making positions. This furthermore aims to identify the potential impacts of the 33.3% gender quota for supervisory and management boards approved in Portugal in August 2017. Consequently, this enables a deeper understanding of the conditions necessary for gender quotas to bring about sustainable gender balances in boardrooms and alongside changes in the corporate culture. This article applies interpretative data analysis from the European Institute for Gender Equality (EIGE) and semi-structured interviews with representatives of nine PSI 20 firms (the 20 largest companies listed on the Portuguese stock exchange). The goals are thus to study the glass-ceiling existence and constraints within the Portuguese corporate context, perceptions on quota outcomes and legitimacy, and the perceived impacts of women on firm performances.

The results of this study demonstrate how quota acceptability is independent of gender as there were no verified gender patterns among the responses obtained; that cultural factors and gender role expectations influence board dynamics and facilitate male access to board positions; and the Portuguese quota requirement may successfully trigger gender parity in the boardroom due to the existence of a sanctioning mechanism for non-compliance and a strong record of path dependency (earlier legal initiatives). However, our findings are inconclusive as regards any possible structural change in corporate mentality and culture as respondents remain divided about the influence of gender diversity on firm performance, the legitimacy of quotas, and their effects on meritocracy.

4.1 Introduction

The fifth United Nations Sustainable Development Goal states the need "to achieve gender equality and empower all women and girls" as gender inequalities remain reflected in many forms around the globe. In western countries, wage gaps, underrepresentation of women in top-decision making positions, and the unequal division of unpaid care and domestic labor constitute the greatest challenges. Overcoming these is fundamental to guaranteeing the sustainable development of societies taking into consideration that gender equality represents an essential pre-condition to economic growth as well as an opportunity in the context of a globalized world still recovering from a deep financial and economic recession. This study focuses on the promotion of gender equality in top decision-making positions, specifically in firms, as women still find themselves underrepresented in the boardroom despite their high qualifications. This not only forms part of an overdue guarantee of equal opportunities for all individuals but also reflects the growing recognition that gender parity is another facet of company Corporate Social Responsibility (CSR) commitments [1].

The specific objective of this article involves identifying the likely impacts of the 33.3% gender quota for supervisory and management boards enacted into law in Portugal in August 2017. This thus also assesses gender discrimination in the Portuguese corporate context, perceptions around the outcomes and legitimacy of quotas, the perceived impacts of women on firm performance, and the efficiency of quotas in promoting board diversity – from the company point of view. Therefore, the main research question approaches whether or not gender quotas may lead to sustainable gender balances in boardrooms and changes in corporate culture. Hence, we test the following hypothesis: as a measure of positive discrimination, quotas may increase the participation of the underrepresented gender in the boardroom even while potentially also raising questions regarding meritocracy, legitimacy, and (female and male) individual acceptance of this form of affirmative action.

The relevance of this study derives from how it tests a specific tool for promoting gender equality – quotas – not only assessing the expected practical outcomes but also debating its ethical dimension. Furthermore, this article responds to the lack of qualitative approaches capable of measuring factors otherwise not perceived by quantitative methods and thus providing a more factual perspective on the topic. The next section presents the literature's findings on quotas thus far – specifically as regard their legitimacy, controversies, and assessed impacts. Section 4.3 then explains the methodology chosen, presents the Portuguese legislation on gender equality, and introduces data on the Portuguese corporate context (macro analysis) before examining the respondent interview answers (micro analysis). The final section both presents the key study results and indicates recommendations for future studies.

4.2 Theoretical Framework

4.2.1 Gender Quotas in the Boardroom

4.2.1.1 Definition

Studies on gender quotas focus on two primary questions: (1) whether or not to apply quotas in certain corporate cultures and, if so, (2) how. The general definition of a gender quota identifies it as a legally prescribed form of affirmative action[1] that sets a percentage objective to implement specific proportional representation of the underrepresented gender – usually women – in management and/or supervising boards of the companies targeted [2, 3]. However, we may furthermore distinguish between *soft quotas* and *hard quotas*. *Soft* quotas amount to benefits attributed to collective entities and organizations for selecting and hiring members of the underrepresented group and thereby serving as a positive stimulus. In turn, *hard* quotas correspond to the legal introduction of a mandatory target for the underrepresented group in boardrooms. In this sense, whenever referring to "quotas," one generally means "hard quotas" corresponding to a legal form of affirmative action [4]. Therefore, and henceforth, this study simply indicates "quotas" whenever referring to *hard* quotas, and "soft quotas" whenever such a distinction is needed.

---

[1] Affirmative action also incorporates the mere indication of individual demographic characteristics that deserve consideration when recruiting (ethnicity, religion, gender, sexual orientation, etc.), without establishing a specific target.

When describing different quota implementation systems, authors usually make recourse to different classifications. However, the most common distinctions stem from the differentiations between **mandatory** and **non-mandatory** quotas (i.e., whenever for reference and voluntarily applied schemes), and among the mandatory, those that imply **sanctions for non-compliance** (with penalties ranging from losing state benefits to stock exchange delisting) and those that do not result in any kind of penalty [5, 6].

4.2.1.2 Quotas: Controversies and dilemmas

As corporate quotas constitute a form of **state intervention** in company dynamics, the debate over their respective legitimacy is prominent in the literature. Equality for all people is a fundamental socio-political goal. Therefore, there is also broad agreement that governments should promote and implement measures to counteract all forms of discrimination [7]. However, in the case of gender quotas in corporations, the state legitimacy for interfering in the governing autonomy of private companies and shareholder choices of board directors has come in for question [8]. The scope for reverse discrimination (against men, usually the overrepresented gender) and the effectiveness of a legal measure directed at a problem that mostly arises out of intricate cultural and sociological roots also contribute to the questioning of state interference through quotas.

There are thus three main controversial topics of discussion surrounding the debate on the legitimacy of quotas: (1) the **justice** of quota implementation [3, 4, 9, 10]; (2) the compromise of **meritocracy** by quotas [3, 7, 8, 10–12]; (3) and the implementation of **work–life balance** policies [3, 7, 13–15].

Regarding the **justice** of implementing quotas, some question their morality. Kogut et al. [2] label quotas a form of state "libertarian paternalism." In terms of the ethical justifications put forward for this, John Rawls *A Theory of Justice* (1971) sets out a theory of social justice that defends two principles of justice: all civil liberties should be guaranteed to all individuals and resources should be distributed in such a way that the less advantaged become the biggest beneficiaries and that equality of opportunities extends through society. The latter is inconceivable without the former and both are indispensable for any ethical evaluation. Thus, Rawls finds quotas morally unjustifiable[2]

---

[2]"(...) the use of soft and hard quotas (...) are – at least from the perspective of Rawls' egalitarian theory – morally unjustifiable. The quota system suspends fair equality

in keeping with how they compromise equality of opportunities in order to guarantee the underrepresented gender holds the right to a seat in the board-room, i.e., the second principle overrules the first. In turn, Robert Nozick's *Libertarian Theory of Justice* rejects state intervention beyond the minimal state (essentially for security reasons) and the social contract although he does consider intervention in the case of historical rectification, i.e., to compensate for disadvantages suffered by groups due to past events that are incoherent to contemporary social standards (e.g., apartheid). However, he does not explain how to implement any such rectification – for instance, through quotas – without sacrificing the premise that when individuals benefit from something acquired legitimately, i.e., due to their own merit, the state must not intervene to redistribute that benefit among others [4].

Overall, support for quotas stems from utilitarian rather than deonto-logical principles with such grounds deeming state intervention necessary, good, and useful to the promotion of equal opportunities rather than simply considered just [4]. Hence, the fundamental role of a company is contributing toward economic and social development through its orientation toward its stakeholders.

This utilitarian logic points to the economic advantages returned by corporate board diversity. This reaches beyond the argument of equality as women get evaluated both as hitherto underrated resources and as sources of important knowledge and experiences. This leads to the *Wollstonecraft Dilemma*: the pursuit of equality between men and women based on their similarities may lead to the measurement of women according to a male standard while arguments based on the differences between the two genders and their benefits to the board may actually disqualify women (because differ-ent characteristics imply different values). The binary thinking of whether to regard both genders as either similar or different represents a debate ongoing in feminist thinking even though there is not necessarily any opposition but rather a constant tension inherent to the meaning of gender [9].[3] Therefore, the pursuit of equality encapsulates the respecting of differences as both con-cepts are not necessarily exclusive. Therefore, are quotas justifiable? Lansing and Chandra [3] and Terjesen and Sealy [10] identify a utility rationale

---

of opportunity in the name of securing equal liberties, but this represents a violation of procedural justice. The use of soft and hard quotas, therefore, cannot guarantee the justness of distributive outcomes, since these outcomes are themselves the consequence of unfair procedures (. . .)" [4].

[3]Characteristics and roles attributed by society and the self whether or not based on one's sex.

seeking to bring about an economically satisfying outcome, and a justice-based rationale that defends fairness, individual, and social equality. Both rationales provide frameworks for arguments both for and against quotas.[4]

We may state that matters of fairness do not only derive from a selection process based on demographics but also include doubts around **meritocracy** [11]. Requiring a certain percentage of the underrepresented gender may compromise selection processes based on qualifications and expertise, even leading to gender bias and to the unnatural selection of qualified candidates from the underrepresented gender, reflecting an over-zealous approach that compromises the normal selection of male candidates [12]. Some authors also warn of the risk of tokenism,[5] reverse-discrimination (against men), the appreciation of gender over competence, cases of nepotism, and triggering opposition to quotas among existing women directors, who may then feel stigmatized [3, 7, 8]. However, some arguments convey how quota systems are compatible with merit-based systems whenever the regulations put in place embody principles of efficiency and complemented by measures promoting networking and mentoring for women directors, as well as work–life conciliation policies [11]. As Terjesen and Sealy (2016, p. 33) [10] note, "despite popular rhetoric, the world is not meritocratic and systemic biases prevent equality of opportunity (...) true merit is only possible in a society without biases."

Another source of opposition against quota systems arises out of the argument that quotas are themselves insufficient when not backed up by structural **work–life balance policies**. These measures aim to provide comprehensive policies, flexibility, and financial conditions for women directors to balance their career goals and family life, especially as regards parental leave (but not exclusively), and also encouraging men to develop other family roles besides that of primary bread-winner – with these policies being independent of gender [7].

---

[4]Examples of arguments of the utility rationale include: "diverse boards may increase financial performance" (argument for quotas); "homophilous boards bring consistency to board processes and decision-making" (argument against quotas).

Examples of arguments of the justice rationale include: "qualified members of the underrepresented gender should have equal access to board seats" (argument for quotas) and "no preferential treatment preventing equality of opportunities should be given to any specific demographic group" (argument against quotas).

[5]Symbolic female representation, i.e., the selection of a few women directors not because of their competences and qualifications but because of their gender [2].

4.2.1.3 Impacts

Regarding the impacts of quotas, some relevant studies in the literature assess and compare the results of quota implementation in those countries where quotas are in place. Most existing studies focus on European countries, particularly Norway, which has the longest period of implementation to date [8, 9, 16], France [17], Germany [7, 13], but also the United States [3]. Most studies study the increase in female directors following the implementation of quotas, which in turn requires more studies on the "quality" of female nominations and the respective processes in order to assess the practical impacts of quotas.

A study based on interviews with board directors in the United States and Europe found that in countries with quotas, there was not only an increase in gender diversity but also a "more professional and formal approach to board selection" [15]. Wang and Kelan [16] found that quota implementation in Norway brought about positive outcomes, increasing gender equality in terms of women holding executive leadership positions while also reporting how women in independent non-executive positions raises the probability of a woman being appointed as board chair. On the contrary, Klettner et al. [18] found, for the same country, that female representation increased following quota implementation but not in executive positions. These authors defend that hard quotas may achieve immediately effective results but that voluntary approaches might achieve greater success in triggering practical cultural and behavior changes.

The number of women on the executive boards of the 200 largest German companies remains only at a very low level even while having risen in the last year of the study, especially at publicly traded companies targeted by the quota and on supervisory boards rather than on executive boards. Furthermore, authors also report a small but statistically significant positive relationship between a higher proportion of female supervisory board members and an increase in their proportion on executive boards [4]. In France, the quota triggered an immediate impact in increasing female representation that Rosenblum and Roithmayr [17] interrelate with the desire of companies to secure beforehand the selection of "the most competent" women directors in the French corporate environment. Directors reported that higher board gender diversity positively affected decision-making and board dynamics and communication not because there were more women but because the new female members represented outsiders to the common elite networks, sometimes foreign nationals, holding degrees and experiences in non-traditional areas, i.e., thanks to all the other levels of diversity they brought in addition to their gender.

As regards the negative impacts, according to Terjesen and Sealy [10], voluntary quota implementation policies that do not include sanctions for non-compliance are less likely to function in practice. In addition, the value of the quota is important to determining the critical mass of women necessary to actually influencing board processes and generating structural equality beyond token representation. Finally, Wyss [19] identifies three pitfalls to the economic rationale that encapsulate hypothetical negative outcome scenarios of quotas: instrumentalism (the selection of women as instruments of profitability, undervaluing their labor), essentialism (exacerbating biological differences between men and women capable of affecting economic outputs whereas the feminist literature highlights differences in socialization and structural inequalities attributing men and women different roles and capabilities), and depoliticizing (the tendency to ignore conflicts between groups and social structures that maintain the existing hierarchies of power).

4.3 Empirical Study

4.3.1 Methodology

We applied a qualitative methodology in order to better understand the expected impacts of gender quotas on promoting management board diversity among listed companies.[6] Specifically, the study objectives involved assessing the extent of gender discrimination within the Portuguese corporate context, the impacts of women on firm performances, and the prevailing perceptions around quota outcomes. Most existing studies attempt to measure board financial performance through quantitative methods, pointing to the lack of qualitative studies in the CSR, gender, and board performance literature. Accordingly, this study adopted a macro approach, through an interpretative data analysis method sourced from the EIGE, and a micro approach, implementing semi-structured interviews, implemented in keeping with our theoretical findings, which we interlinked with the categorized and coded answers of respondents. Both approaches are duly set out in the following sections.

---

[6]Portuguese law No. 62/2017 targets management and supervisory boards of public and listed companies. However, the focus of this empirical study includes only the management boards (because these usually gain greater visibility within companies) of listed companies (because these represent a specific group of firms that choose a highly regulated public mechanism to obtain financing).

Interviews allow for the in-depth gathering of different perceptions, unexpected factors, and holistic depictions, examining phenomena and actors in their own environments. In the case of board dynamics, these are essential to assessing the practical effects of diversity that depend on individual relationships, corporate culture, and decision-making processes – factors not susceptible to reduction to a few variables through quantitative methods [20, 21]. The objective here was to build up a comprehensive understanding of the topics under study by letting interviewees speak freely about their experiences. Regarding content analysis, the questions were prepared and organized in keeping with the main outcomes of the existing literature with the answers then coded and categorized accordingly.

PSI 20 firms were selected for interview due to their dimension and representativeness in the Portuguese economy. Interviews were conducted with representatives from the following companies: *BCP, CTT, Galp Energia, Novabase, Pharol, REN*, a Portuguese energy sector company, and two companies that wished to remain anonymous. The interviews lasted for a duration of 30–40 min and, except for one phone interview, took place face-to-face. *Altri, Corticeira Amorim, EDP Renováveis, Ibersol, Jerónimo Martins, Montepio, Mota-Engil, NOS, the Navigator Company* and *Sonae Capital* did not participate in this study either because the company was not available; because it did not answer the request for an interview; or it was still monitoring the consequences of the legislative proposal. Participants were members of human resource (four participants) and sustainability departments (three participants). Furthermore, one participant held the position of CFO and another served as CEO.[7] The participant age distribution was the following: two aged between 25 and 34, three between 35 and 44 years old, and one between 45 and 54 years old with the three remaining aged between 55 and 64. Table 4.1 details the codes attributed to the respondents.

The interview script consisted of four groups of retrospective and prospective questions with each focused on a specific topic following a method of idea concentration through thematic analysis. Group I "General Perspectives" focused on interviewee ideas on gender equality and gender characteristics; group II "Portuguese Legislation" examined their views on quotas as a mechanism for promoting gender equality and, specifically, the terms of the Portuguese quota; group III "Quota Impacts" assessed company

---

[7]Following a uniformity criterion, the identities of all interviewees remain anonymous. Respondent departments or positions held in the firm and their ages were not associated to each individual in order not to compromise their identity.

Table 4.1   Interview participants codification

Interview Participants		
Respondents	Gender	Interview Recorded
Respondent A	Male	Yes
Respondent B	Female	Yes
Respondent C	Male	No
Respondent D	Female	No
Respondent E	Female	Yes
Respondent F	Male	Yes
Respondent G	Female	Yes
Respondent H	Male	No
Respondent I	Female	Yes

expectations of quota implementation; and group IV "Conciliation Measures" investigated work–life balance measures in place in the companies under study.[8]

Groups I and II are necessarily directed at the personal views of participants, significant as these interconnect deeply with the personal experiences of interviewees in the organizations in which they work. There is a gender-balanced representation among respondents, thus allowing for non-gender-biased answers to these groups of questions. Groups III and IV target their company positions toward the Portuguese legislation on quota implementation, revealing a glimpse of the receptivity of quotas among the firms targeted.

4.3.2 Portuguese Context

In Portugal, women make up 52.6% of the population. According to Pordata, in 2015, 59% of graduates were women while 63% of all individuals holding Master's degrees, and 54% of those holding Ph.D.'s were female. Nevertheless, women still remain underrepresented in all senior corporate decision-making positions in Portugal, even though there has been a slight increase in recent years, as detailed in Figure 4.1. Appendix II presents information about the portuguese position in the Gender Equality Index.

---

[8]Almost all company representatives interviewed mentioned some initiatives run by their firms in this area, which demonstrates how these organizations are not indifferent to this issue. Each firm's initiatives reflect different approaches and the specific characteristics of each sector. However, company annual and sustainability reports rarely include specific chapters on work–life balance and, except for a few respondents, participants provided general answers to this group of questions. For this reason, we decided to analyze this topic in future studies.

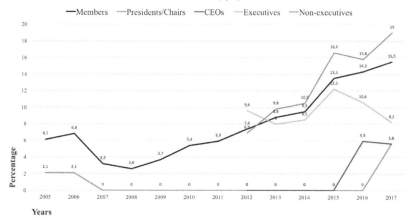

Figure 4.1 Women's representation in the largest listed companies in Portugal, 2005–2017, (%).

Source: Gender Statistics Database, EIGE. Data analysis undertaken by the author.

The lowest level of female board member representation in the highest decision-making bodies of the largest listed companies in Portugal came in 2008 (only 2.6%). Since then, female representation has experienced steady growth and, in 2017, 15.5% of board members were women. Up until 2012, there are no records of any female executive or non-executive members in the two highest decision-making bodies of each company. Only in 2016, were the first female CEOs registered, and with the first female presidents taking office in 2017, after a long period without female presidents. Nevertheless, the proportions depicted in Figure 4.1 remain extremely low in relative terms and the reason Table 4.2 proves helpful in demystifying real female representation on boards.

Table 4.2 presents the main characteristics of PSI 20 company (the largest companies publicly listed on the Portuguese stock exchange) board members, specifically average director age, qualifications, and independence and especially focusing on gender. PSI 20 companies are individually very different from each other, depending on their governance model and sector, which shape their corporate cultures and governance strategies. In particular, board sizes vary significantly (from three members in the case of *Ibersol* to 21 members in the case of *BCP*), which is meaningful when analyzing female representation. In some cases, adding one or two women is enough

Table 4.2 Characteristics of PSI 20 firms and descriptive statistics

Characteristics of PSI 20 Firms and Descriptive Statistics, 2016

PSI 20 Firms	Board Size	Average Age	♀ Directors	♀ Board Representation (%)	Executive Members	♀ Executive Members	♀ Executive Members (%)	♀ Board Chair	♀ CEO	CEO Duality[a]	Independent Non-Executive Members	Bachelors[b]	Masters; Ph.D.[b,c]	MBA; Executive Training; Post-Grad.[b,d]
Altri	7	N/A	2	28.57%	4	0	0.00%	No	No	Yes	0	7	0	3
BCP	21	57.9	3	14.29%	8	1	12.50%	No	No	No	7	21	5	6
Corticeira Amorim	6	52.8	2	33.33%	3	0	0.00%	No	No	Yes	0	5	0	3
CTT	13	54.6 e)	4	30.77%	5	1	20.00%	No	No	No	6	12	4	3
EDP[e]	8	53.4	0	0.00%	8	0	0.00%	No	No	f	8[g]	7	2	6
EDP Renováveis	17	57.7	1	5.88%	4	0	0.00%	No	No	No	10	17	11	7
Galp Energia	19	55.6	3	15.79%	7	0	0.00%	Yes	No	No	5	19	7	8
Ibersol	3	N/A	0	0.00%	2	0	0.00%	No	No	Yes	1	3	1	2
Jerónimo Martins	9	59.9	1	11.11%	1	0	0.00%	No	No	Yes	5	8	4	3
Montepio[e]	7	N/A	0	0.00%	7	0	0.00%	No	No	f	0[g]	7	1	4
Mota-Engil	17	N/A	3	17.65%	8	0	0.00%	No	No	No	3	17	4	3
NOS	17	N/A	4	23.53%	7	1	14.29%	No	No	No	1	14	2	11
Novabase	4	52	0	0.00%	2	0	0.00%	No	No	Yes	0	4	2	2
Pharol	10	57.4	1	10.00%	2	0	0.00%	No	No	Yes	3	10	2	6
REN	12	52.3	2	16.67%	3	0	0.00%	No	No	Yes	4	11	4	5
Semapa	11	N/A	0	0.00%	4	0	0.00%	No	No	No	3	10	1	3
Sonae	9	59	3	33.33%	2	0	0.00%	No	No	Sim	7	9	5	5
Sonae Capital	7	54.1	2	28.57%	3	2	66.67%	No	Yes	No	2	7	1	6

The Navigator Company	14	N/A	0	0.00%	5	0	0.00%	No	No	No	0	13	1	8
Mean	11.1	55.3[h]	1.6	14.18%	4.5	0.3	5.97%	–	–	–	3.4	95%	26%	48%
Median	10.0	55.5[h]	2.0	14.29%	4.0	0.0	0.00%	–	–	–	3.0	100%	24%	50%
Mode	7.0	51[h]	0.0	0.00%	2.0	0.0	0.00%	–	–	–	0.0	100%	0%	50%
Standard Deviation	5.1	8.8[h]	1.4	12.21%	2.3	0.5	15.44%	–	–	–	3.0	6%	18%	18%
Max.	21	81[h]	4	33.33%	8	2	66.67%	–	–	–	10	100%	65%	86%
Min.	3	35[h]	0	0.00%	1	0	0.00%	–	–	–	0	82%	0%	18%

♀, female

N/A, not available

[a]The CEO is also the chairman of the board of directors.

[b]Descriptive statistics correspond to the proportion in percentage of degrees in each board.

[c]When an individual holds both a Master's degree and a Ph.D., only one qualification is counted.

[d]MBA, executive training, post-graduate qualifications. When an individual has completed more than one type of program, only one is counted.

[e]The age of one member is not available.

[f]Dual model (executive board of directors, general supervisory board and statutory auditor).

[g]Independent members of the executive board of directors.

[h]This accounts for the ages of all individual members of each company, and not the average age in each company as indicated above.

Source: Official company annual reports (publicly available on their websites). Data analysis by the author.

to significantly increase female representation even while this does not apply in the case of larger boards. Even for smaller boards, compliance with the gender proportions required by law may still lack significant female influence. The median of female board directors is two and thus the median (and as well as the mean) female representation in PSI 20 firms is only 14%. The firms returning the highest levels of female board representation are *Altri* (28.57%), *Sonae Capital* (28.57%), *CTT* (30.77%), *Sonae* (33.33%), and *Corticeira Amorim* (33.33%). The latter two are the only companies that, at least in terms of the board of directors, comply with the new Portuguese quota law. However, there are no women sitting on most executive commissions. This means that even the few women holding board seats do not attain executive positions and, therefore, have only a reduced influence on decision-making relative to their male peers even though independent non-executive female directors take on important board of director supervisory roles. *Sonae Capital* constitutes the only exception to this scenario: out of its three executive commission members, two are women and one of these is the CEO – the only case of a female CEO in PSI 20 firms. Furthermore, *Galp* represents the only case of a female board chair. Analysis of female board representation must take into account the distinction between executive and non-executive members as the increase in female board representation may result only from increasing female non-executive members, reflecting tokenism [18, 22]. As predicted by Nekhili and Gatfaoui [23], there is no clear evidence that boards with a higher number of non-executive directors are more likely to appoint female directors.

The average age of directors for all PSI 20 firms is 55 years. In *Sonae* and *Jerónimo Martins*, the average age is higher (59 and 59.9, respectively), and lower in *Novabase* and *REN* (52 and 52.3, respectively). Nowadays, the board of directors belongs to Generation X when women began fully accessing academic education and hence enabling an equal and fair assessment for the levels of qualifications attained. Therefore, we may also correspondingly expect that, other things being equal, today's directors would be representative of both genders. Indeed, when analyzing director qualifications, there are no significant differences between men and women. Globally, and apart from a couple of exceptions, all directors have undergraduate degrees, some have Master's degrees, a few have Ph.D.'s, and many have MBAs or other types of executive qualifications (usually more than one). This suggests the reasons why women are not reaching the boardroom arise out of other factors.

4.3.3 Legislative Framework in Portugal

In Portugal, the legislative framework establishing gender equality stems from, first and foremost, Article 9, paragraph h) of the Constitution, which defines this as one of the fundamental duties of the state through the prevention of gender discrimination and the promotion of affirmative action. As a result, there are legal mechanisms that, in recent years, have given rise to a relatively consistent – although not actually effective – framework supporting gender balance in the workplace, specifically within firms.[9]

Pointing to the insufficiency and ineffectiveness of past voluntary auto-regulatory mechanisms and the persistence of gender inequalities in corporations, the Portuguese government proposed draft law No. 52/XIII, "establishing the regime of balanced representation of men and women in management and supervisory boards[10] of public sector organizations and listed companies," on January 5th, 2017, which was then approved and enacted by law No. 62/2017 on August 1st, 2017. The initiative seeks to promote gender equality in senior decision-making positions and forms part of a set of government initiatives for gender equality within the overall objective of generating sustainable economic growth, an efficient use of qualifications and competences, and a more just and inclusive society.

Law No. 62/2017 stipulates that the proportion of each gender sitting on management and supervisory boards (including executive and non-executive directors) of public companies can be no lower than 33.3% as from January 1st, 2018. For listed companies, this cannot be less than 20% as from the first elective general assembly held after January 1st, 2018, and 33.3% as from the first elective general assembly after January 1st, 2020. Non-compliance by public companies results in the invalidity of the board's nomination and a 90-day period for a new nomination. In the case of listed companies, the CMVM – the Portuguese Securities Market Commission – first issues a declaration reporting non-compliance before a 90-day period to change the

---

[9]See relevant legislative measures in Appendix I.

[10]According to the Portuguese Company Code, article 278, there are three models of corporate governance for companies, regarding their management and supervisory bodies:

    a) **Latin single model:** board of directors (includes an executive commission) and fiscal council;

    b) **Anglo–Saxon single model:** board of directors (includes an executive commission), audit commission, and statutory auditor;

    c) **Dual model:** Executive board of directors, general supervisory board, and statutory auditor.

board's appointment after which, should non-compliance persist, a reprimand is then publicly handed down. Whenever the non-compliance still persists 360 days after the reprimand, the company becomes liable for the payment of a mandatory fine (not more than a month of salaries of board members for each semester of non-compliance). Law No. 62/2017 also defines how listed companies ought to present annual plans for promoting both the equal treatment of genders and the reconciliation between personal and professional lives.

Terjesen et al. [6] investigated the factors that may drive governments to legislate over gender quotas for corporate boards. The authors propose a model featuring three institutional dimensions that raise the likelihood of recourse to gender quotas: the female labor market and the pre-existence of gendered welfare policies targeting female employers and conciliating working and family life, left-wing government coalitions, and a historical record of path dependency (legal initiatives for gender equality). Within this framework, even non-binding mechanisms, such as codes of corporate governance and stakeholder expectations, can foster strong normative pressures. These result in a certain mimetic isomorphism, i.e., due to the incapacity to find solutions for uncertain conditions, firms choose to implement already existing systems (e.g., Norway), generating standard answers and a general institutional change. In Portugal, we may verify all three components although, as identified by technical opinions from civil society entities,[11] there is a lack of structural measures capable of supporting the quota system, for example promoting wage equality and work–life balance measures that particularly target women.

4.3.4 Analysis of the Interview Results

4.3.4.1 Perceptions of gender equality

First, we assessed to what extent interviewees perceived gender discrimination and the existence of the glass ceiling for senior management positions as a reality in Portuguese corporate culture. Every participant, except for Respondent C, accepted the existence of a glass ceiling globally although they all defended that this did not represent the current reality at the companies for which they work. Respondents E, H, and I pointed out that unequal gender representation very much depends on the company and its sector, with Respondent I identifying a bottleneck around the senior leadership

---

[11]Particularly those published by the National Confederation of Portuguese Workers (CGTP-IN).

positions in her firm. Furthermore, when confronted by the figures portraying the low female representation on PSI 20 company boards, all respondents presented several justifying factors. These fell into three categories: **firm-specific**, **societal/cultural**, and **gender-related**. Participants usually referred to more than one type of argument in their answers.

Firm-specific arguments evoke the rights and the authority of company shareholders to select board members. Respondents A, C, D, F, and H explained that as board nominations are made by shareholders, gender parity does not depend either on the company or on its culture but on shareholder judgments of the fittest and most trustworthy candidates for directorships.

> "(...) male and female directors are always selected or nominated by shareholders, therefore the board of directors ends up not having that freedom [of selecting its directors] (...)"

> – Respondent F

Societal/cultural factors were those most referenced by respondents through abstract depictions of Portuguese demographics and societal values they believe get reflected in the make-up of company boards. Respondents B, D, E, G, H, and I mentioned the different roles traditional Portuguese society attributes to the respective genders: women as housewives and primary care-givers and men as bread-winners. According to the views of participants B and D, these roles not only shape expectations toward members of both genders but also influence behaviors in the workplace. Men thus become expected to remain in the office after work to engage in informal gatherings that usually both discuss promotions and nurture business networks. Women rarely get to participate as they usually leave the office to head home and do the housework and take care of their children (commonly responsibilities not shared with their husbands) and thus seem less available for work:

> "(...) men interact more with each other, women go home at the end of the day, (...) because they have to take care of their other tasks and will not go out for drinks (...). Usually, it is in these moments that people are chosen and potential successors are considered (...) the model is indeed dominated ... by the masculine model and by men. And it's natural that they choose men because they interact more with each other outside of work."

> – Respondent B

Moreover, Respondent C pointed out that today's directors belong to a generation in which more men than women entered the corporate circles and that boards are tending to diversify overtime. Accordingly, Respondent I believed that over time the role of "primary care-giver" attributed to women will become independent from gender.

Four respondents (B, D, H, and I) referred, directly or indirectly, to **gender-related factors**, in the sense that most men seem to be members of relevant networks ("boys' clubs"), from which women are somewhat excluded, thereby favoring male candidates.

> "It's natural that they [men] tend to choose people from their circle of confidents."

– Respondent D

Respondent I identified a corporate habit of neither including nor integrating women into discussions and decision-making as contributing to weak female representation:

> "Decision-making takes place within mostly male groups (...). In equal circumstances and following a logic of meritocracy (...) maybe, sometimes, decisions are made because we do not ask women if they are available"

– Respondent I

Respondent H mentioned that directors look for successors through a process of identification with themselves and that gender roles also interrelate with leadership characteristics attributed to each gender, especially assertiveness and aggressiveness in men, which are preferred in competitive environments. Regarding the female gender, Respondent E stated that women are often the first ones to limit themselves by being overly insecure about their capabilities and less prone to risk-taking than men:

> "Many times, women are their own constrainers, that is, when there's an opportunity for career development, it is women who usually choose not to take the risk, because they tend to (...) better preserve the feeling of security (...). Women are less confident about their competences than men but which doesn't amount to a rule [for all women]."

– Respondent E

4.3.4.2 Perceptions of gender diversity impacts

Next, we then examined considerations around the impact of women on firm performance levels. This question contained two focuses: **financial** performance and **non-financial performance**. However, all interviewees initially mentioned factors related to **non-financial** performance in their answers, and only interviewees C and F did not think gender diversity could have any impact on firm performances. Most interviewees (A, D, E, G, H, and I) believed that diversity in all of the individual characteristics, their points of view, and backgrounds is susceptible to affecting decision-making, risk evaluation, and strategic thinking, and not only gender diversity:

> "There are different ways of seeing the world and life that don't come only from gender, they come from gender, from what we are as individuals, from where we grew up, what our social context is, our economic contexts, academic contexts, family contexts... all of this influences the way we see the world, not just the gender question, but that also has an influence."
>
> – Respondent A

Thus, when talking about outcomes, respondents distanced themselves from gender and with the most common argument being "competences are independent from gender." Respondent E, however, stated a belief in more women in the boardroom contributing toward a more positive and ethical company image, which thereby increases shareholder confidence and, therefore, investment, due to women being associated with more ethical attitudes. Similarly, Respondent B believed that female directors usually attribute a greater weighting toward sustainability-related topics and boost their development within the organization. This opinion contrasted with that of Respondent G who believed male and female sustainability directors share equal concerns and that approaches depend more on the company's sector and its impact on the environment and stakeholders than on gender.

> "(...) those assumptions of women being more closely associated with a certain kind of concern are wrong (...). It is actually very dangerous to assume those positions (...) that in reality are myths (...). There are very few female CEOs of oil and gas companies, for instance, but this does not mean that the sustainability concerns of those companies are less relevant."
>
> – Respondent G

When asked about **financial** performance, interviewees tended to repeat themselves and refer to the benefits of diversity in decision-making. Only Respondent D said that as female graduates are leaving university with higher grades relative to their male colleagues, hiring female talent is "good business" for companies, which also holds true for women in senior positions aiming for the board of directors. Similarly, Respondent E mentioned that organizations are now becoming aware of this and moving to increase female representation due to competition with other firms for new and diverse talent.

> "We always want to be the best of the best (...) and the question of diversity and women in top management positions has become something perceived as positive for companies, something that will generate a return."
>
> – Respondent E

4.3.4.3 Views on quotas

Subsequently, the interviews focused on the topic of quotas, their acceptability, and expected impacts. Even those favorable to quotas expressed preference for other approaches to affirmative action and specifically mentioning company self-regulation (along with public scrutiny), fiscal incentives, and rewards, raising awareness and work–life balance initiatives. Only three respondents **agreed** with the implementation of quotas (B, E, and H), with three **against** (C, G, and I), and with three believing that taking into account the present scenario, they should be implemented, despite **not being an ideal option** (A, D, and F). Follow-up questions tried to grasp the extent to which the respondent positions arose out of the three main ethical dilemmas quotas set out in point 1.1.2: **reverse discrimination, meritocracy,** and **state intervention**.

Individuals in favor of quotas affirmed strong beliefs as to how they might trigger behavior change due to their mandatory character and sanctions. They focused on the fact that, contrary to what one might otherwise think, quotas amount to more than a positive discrimination measure:

> "I think quotas can be really useful (...), not as a concept of positive discrimination but in the sense of equality, of giving equal opportunities (...)."
>
> – Respondent B

In Respondent E's understanding, the promotion of equal opportunities through quotas comes with rigorous selection criteria in order to prioritize overall candidate qualifications. Only after having secured these should applications by the underrepresented gender be favored:

> "(...) it's about equality of opportunities, while always referring that, for equal talent, the underrepresented sex does benefit but always highlighting that the level of talent has to be the same (...)."

> – Respondent E

Therefore, respondents rejected **reverse discrimination**. Respondent H termed the reverse discrimination argument "dishonest" and stated that unless we actually believe men are better at doing things, we cannot explain how among the top 100 there are 90 men. This clashes with the opinions of Respondents C and I who deployed the argument of **meritocracy**, but in reverse, explaining that quotas may compromise merit and prevent the most qualified candidates from being appointed if male.

> "(...) quotas are not sustainable in the long run, otherwise, firms would easily recruit non-executive directors (...) and the matter would be solved. (...) It is not about gender equality but equality of opportunities (...) that's why we defend merit."

> – Respondent I

Cultural factors were also applied reversely with Respondent G stating that such a mechanism makes more sense in highly conservative societies in which women do not enjoy the same civil liberties as men. Contrarily, Respondent E said that our gender-role-based-society justifies the existence of quotas.

> "[In a] geographic area (...) where, in cultural terms, women do not have a role... in terms of the labor market (...) the existence of quotas may be a [necessary] mechanism to create a culture of diversity in those organizations."

> – Respondent G

> "In Portugal, Spain, highly traditional, catholic countries, there are specific gender roles, and maybe, in Northern countries, there is not such a strong historical and religious component (...). Nowadays, we are in a more and more globalized world (...) and those cultural issues (...) have lesser importance."

> – Respondent E

Those who believe quotas are not the ideal option for promoting gender equality understand the **state's intervention** as stemming from the urgent need to change behaviors. They view quotas as a temporary mechanism and therefore reject the idea they may **compromise merit** or **discriminate against men**.

> "(. . .) I do not see these [quotas] as negative discrimination for men because I believe they are temporary, and, if this practice becomes generalized, quotas will stop making sense."
>
> – Respondent A

Specifically regarding the **Portuguese quota**,[12] only Respondents A and G came out in agreement with the overall proposal, particularly with both the percentage and the defined implementation period. Respondent C, however, strongly opposed the quota, considering it hastily proposed and resulting in one of the strictest frameworks in Europe. Respondents F and H believed the distinction between listed and unlisted companies as targets of the measure represents an additional burden for the former before proposing a dimension criterion as more appropriate. Respondent E questioned the reasons behind the quota percentage, recalling how 33.3% represents only one woman on smaller boards, which may translate into tokenism and the subcriticality of women. Respondents B, C, D, and I identified this goal as "ambitious" and potentially representing a "challenge" for most of the companies targeted. However, all company representatives guaranteed their organizations would be able to implement the quota but with only the anonymous firm and the energy sector company mentioning a bottom-up approach involving the broadening of their talent pool, investing in training their human resources from the start of their careers, and working on eliminating internal bias in career development and decision-making.

Afterward, we attempted to understand the **expected impacts** of the Portuguese boardroom quota system in terms of fostering promotion of opportunities among qualified senior females, breaking the glass ceiling, and improving decision-making dynamics. However, when asked about the feasible impacts, interviewees tended to repeat themselves and give similar answers to when asked about the impacts of women on financial and non-financial performance. After some follow-up questions, no one reported more women in the boardroom as likely having any specific impacts beyond representing additional motivation for younger female professionals:

---

[12] Some of the interviews were made before the draft law approval.

"I think women "attract" women, especially in management. When there's the perception that that person achieves a certain position, there are more people showing up for that position, because, suddenly, the dream comes true (...)."

– Respondent E

Respondents A and B **did not expect any impact**. Respondent A mentioned gender leadership characteristics to argue that, from his own experience, these are mere stereotypes and individuals usually do not conform with them, giving the example of the belief that some women directors have a more emotional and sensitive – i.e., feminine – approach, and others a more aggressive – i.e., masculine – posture.

"(...) [when we talk about] women who manage in a little more 'masculine' way, a little less relational – even though here we are talking about stereotypes too – we're supposing that women manage in a more emotional way than men (...); during my career (...) I've worked with people that do not fit these typologies (...)"

– Respondent A

Respondents E and G referred to **impacts in terms of diversity in general** (across all its dimensions), and not just gender diversity, as an advantage in decision-making.

"I consider (...) diversity as a competitive advantage (...), it prepares boards more adequately for decision-making, which is, obviously, positive in that perspective."

– Respondent G

Respondents C, D, H, and I referred to some **specific impacts of the quota** measure. Respondent C expressed concern over the changes necessary to executive commissions because each executive member would now have to hold a specific portfolio. Hence, either some portfolios would have to be split, or more created, or male members would have to leave in order to integrate women directors. Respondent D mentioned improvements to company public images and the social impact as positive outcomes of the measure alongside Respondent I who added that by valuing and investing in diversity, firms contribute to changing mentalities and nurturing their internal and external images, reflecting the Perrault [24] and Rao and Tilt [25] findings. Respondent H referred to the benefits of diversity in general but, above all, diversity among women themselves, i.e., as women begin accessing senior positions, they gain the opportunity to develop individual competences and become valued for them and not for gender-related aspects.

Nevertheless, as no answer referred to **negative aspects**, when confronted with this possibility, participants had much more to say, particularly regarding follow-up questions on **tokenism** and **meritocracy**, and usually mentioning more than one type of impact. Respondents A, H, and I denied the possibility of negative impacts. Respondent I mentioned, briefly, some eventual difficulties in managing change and reorganizing teams.

> "(...) maybe performance will decline temporarily a little, and then increase significantly, but that happens in all learning processes."

> – Respondent I

Respondent A mentioned that the likelihood of selecting **unqualified members** remains questionable as both men and women have similar levels of qualifications following the access of women to education becoming a reality many years ago. Furthermore, even if there were signs of **tokenism** in the early years of implementation, Respondent H did not believe this would compromise the positive impacts of the measure. Respondents D, E, and F, however, mentioned **tokenism** as a possibility. Respondent D explained that were women to feel favored because of their gender, this might pose a real obstacle to their performance. Respondent E pointed to the need for a critical mass of women in order to avoid tokenism and difficulties in changing the mindset of workers in more conservative companies that perceive quotas as unfair positive discrimination in favor of women:

> "We need time to explain people the reasons for supporting quotas (...), and sometimes, that is time companies do not want to waste, (...) it is difficult to work in a culture with which people can't identify (...), we would be evangelizing people (...). People will always think: 'oh so we're getting rid of good men to give women their positions' – and that is not true."

> – Respondent E

Respondent F predicts the appearance of **Golden Skirts**[13] and only Respondent C expressed concerns over **meritocracy** and either the selection of less qualified people or the non-rewarding of those most qualified.

---

[13]Qualified women directors who hold several directorships on a number of different boards [3].

"(...) there will be women in board cycles and they will end up being in several places [boardrooms], and sometimes they will not be chosen on merit. They will be chosen just because they are women and have already held that position so they can do it again in other places."

– Respondent F

Respondent G mentioned negative impacts of the measure to the public images of companies as quotas may lead to the false impression that companies are only promoting female representation because it became mandatory by law. This does not fit in with any of the categories above and also contrasts with the opinions of Respondents D and I who perceived the quota as resulting in positive impacts and improvements to company images.

"This gives the wrong idea about the concerns of organizations as regards equality of opportunities because it gives the false idea that we only do so out of a matter of legal obligation."

– Respondent G

"(...) to lead by example in that matter, so those at the bottom understand that it is possible to diversify and that it [diversity] is valued."

– Respondent I

4.3.5 Discussion of Results

As predicted by Terjesen et al. [6], at the time of approving the quota legislation, Portugal had all three components in place for implementation, in particular, a left-wing government coalition and a record of path dependency of previous measures in this field (Appendix I). However, implementing a 33.3% quota by 2018 does represent a significant challenge for the largest listed companies. Although guaranteeing they will be able to meet the quota, the interviews revealed how some firms are not favorable to the mechanism chosen. A slightly longer period of implementation, of 2–6 years, similar to other European countries, might not only have facilitated adjustment to the quota but also have allowed for raising awareness among all the parties involved. This is important because, on some boards, achieving a 33.3% proportion may imply the nomination of only one woman. Ideally, female nominations should not be stigmatized as the token women or the "Other," although this may be initially a risk as female director competences may get overlooked because of their gender.

Overall, respondents were quite open to discussing the topics under study. However, most encountered difficulties in conveying the formal positions of their firms on quota legislation. They instead mostly expressed their personal opinions and experiences, even while still valid given these reflect the realities of their firms. It would thus seem that there may yet not have been substantial internal discussion, which raises concerns due to the tight timeline for implementing the quota. This aspect along with just which companies to target and the appropriate level of the quota (to avoid tokenism and subcriticality) formed the doubts raised by respondents. Specifically, some respondents suggested that public and state administrated companies should be the first to set the example, and that quotas would harm listed companies by imposing an additional burden. Moreover, some defended that a dimension criterion would have been more adequate as some listed companies are small/family-run firms. Regarding the quota type, many were in favor of soft quotas, especially self-regulatory approaches, while others stated that, taking into account the present context, hard quotas are a necessary instrument for driving change. This in turn encapsulates the different conceptions surrounding the best way of triggering cultural and behavioral changes and the controversies existing between the hard and soft approaches [18].

In light of the results obtained, the interviews reveal some important and noteworthy patterns: **(a)** as described above, gender did not hold any clear influence on the acceptability of quotas (two women and one man were in favor and two women and one man were against). This instead depended more on respondent experiences, company backgrounds, company cultures, and, ultimately, individual values; **(b)** in terms of age, there were no clear differences among the answers given; **(c)** female director seniority did make a difference: females with long careers within their companies were usually against quotas and seemed protective of their status quo. They usually held top positions and stated they would not like to benefit because of their gender because this would devalue their individual competences. This conveys the risk of female director opposition to quotas as predicted in the literature; **(d)** some respondents note that sometimes women impose obstacles on themselves: they tend to be less aggressive and less participative in stating their suggestions and in decision-making, they hesitate when an opportunity or challenge emerges, and they appear insecure as to their own competences and abilities to play a businessperson role alongside that of mother/wife.

The literature reports this as a generalized remark in opposition to supposed greater assertiveness of males [3].

Regarding respondent perceptions of gender equality, "societal/cultural arguments" constituted the most mentioned facet and, nowadays, seem to reflect a long overdue demand of second-wave feminism that, in the late 1960s, had questioned gender roles and the division of tasks in western households. In turn, "gender-related" answers identified – directly or indirectly – homophily,[14] power relations, gender stereotypes unfavorable to women, and the so-called "boys' clubs" as responsible for excluding women from eligible positions, relevant networks, and favoring male candidates.

Concerning the impacts of gender diversity, most interviewees considered competences as independent from gender and did not allow for any positive relationship between gender diversity and financial performance. However, most also considered this beneficial to non-financial performance, decision-making processes, and board dynamics as a trait of diversity, and not specifically because of gender.

Nevertheless, the legitimacy of quotas remains controversial and an ongoing debate. The antagonistic views on reverse discrimination, tokenism, golden skirts, and meritocracy mentioned by respondents emerge in the findings presented. This reflects the current debate among researchers and corporate workers as regards the following topics: the scope for quotas to represent an unfair positive discrimination measure favoring women and leaving men in a disadvantaged position; the risk of female directors being appointed not for their competences but for symbolic purposes (and of getting appointed to more than one directorship); and the compromising of merit in selection processes. These topics were broadly present in different respondent answers. Furthermore, the opinions of Respondents D and I, that boardroom gender diversity improves company public and internal images as well as worker trust, reflects the Perrault [24] and Rao and Tilt [25] findings. Contrarily, Respondent G defended that corporations implementing quotas encapsulate the idea that they only try to promote gender balance because this is mandatory under the law.

Respondents made important comments on adapting to the measure with some concrete suggestions: businesswomen ought to be very clear and assertive concerning their professional objectives, salaries, and interpersonal relations in the company. In turn, companies themselves have to guarantee

---

[14]A board's particular trait, visible when its members present similar characteristics, including gender [24].

rigorous selection criteria and explain the importance of quotas to their workers and how, if well implemented, they will not compromise merit; the role of education in promoting gender equality, deconstructing stereotyped frameworks, and shaping positive and sustainable social behaviors in order for quotas to be unnecessary in the future.

4.4 Conclusion

This article's main objective was to identify the likely impacts of the 33.3% gender quota for management boards in Portugal in order to understand the terms on which gender quotas may or may not trigger sustainable gender balances in the boardroom and change in the corporate culture. This work deployed data analysis and semi-structured interviews to investigate the existence and constraints of the glass ceiling in Portuguese corporate culture, perceptions on quota outcomes, and the perceived impacts of women on firm performance levels. Examining this topic from the perspective of organizations made it possible to assess the receptiveness to the quota system that come into effect in 2018.

The starting hypotheses suggested that quotas may increase the boardroom participation of the underrepresented gender but while also raising legitimacy concerns. Globally, the interviews approached this dilemma positively. However, the literature review and data analysis suggested a more complex answer to the research question. The success of any quota system deeply depends on its formulated terms, on a country's corporate culture, on social receptivity and, at the micro level, on the sector an organization belongs to. Moreover, cultural aspects, ethical values, and views on meritocracy hold great influence over corporate social dynamics and over the acceptance of quota as testified to in our interviews. Furthermore, views on these topics divided broadly. Participants attributed great weighting to society's instituted gender role expectations. The interviews demonstrate how quota acceptability is independent from gender as are perceptions around the benefits of gender diversity. Thus, quota acceptability varies according to the personal values, experiences, and corporate backgrounds of individuals. Correspondingly, although some respondents recognized benefits for board dynamics from the different traits inherent to both genders, others strongly denied gender diversity having any impact on firm financial performances.

Regarding the main objective of this study, we may conclude that, in the Portuguese case, its record of path dependency, i.e., earlier legal initiatives, is significant to the success of the quota system. Previous soft approaches established the necessary impulse for hard quota implementation although some technical and legal opinions have criticized the legal quota framework. The quota system will most likely guarantee a gender balance in the boardroom at a sustainable level. Nevertheless, we are unable to ascertain whether or not this will trigger change in corporate mentalities and cultures as respondent answers were divided. However, data analysis made it clear that the glass ceiling is a contemporary reality to the Portuguese corporate context.

There are some recognized limitations to this analysis. Despite several interview requests, some PSI 20 firms did not participate in this study, which decreased its potential, as the interview method requires a varied number of testimonials. Additionally, the firm representatives interviewed were neither all board members nor members of the same departments, which invariably results in different perspectives on the topic under research. Furthermore, interviewer subjectivity and inaccuracies in the interpretations of answers pose risks although we asked open-ended questions in order to minimize answer biases. Moreover, the impacts of gender diversity on firm financial and social performances might have been further developed. In particular, future studies should strictly define the dimensions to performance and test for the impacts of gender diversity through qualitative methods in order to measure the actual impacts of the quota after a certain distance in time following implementation. Given this study assesses the potential of hard quotas as a gender equality instrument, it is important to examine their actual outcomes, similar to quantitative studies on the Norwegian case and other countries with longer established quota systems. However, the introduction of qualitative methods in a later phase would be equally relevant in order both to overcome the limitations of quantitative methods and to gather the hitherto unperceived aspects of quota implementation.

References

[1] Grosser, K. (2016). Corporate social responsibility and multi-stakeholder governance: pluralism, feminist perspectives and women's NGOs. *J. Bus. Ethics* 137, 65–81.

[2] Kogut, B., Colomer, J., and Belinky, M. (2014). Structural equality at the top of the corporation: mandated quotas for women directors. *Strateg. Manage. J.* 35, 891–902.

[3] Lansing, P., and Chandra, S. (2012). Quota systems as a means to promote women into corporate boardrooms. *Employee Relat. Law J.* 38, 3–15.

[4] Hall, S., and Woermann, M. (2014). From inequality to equality: evaluating normative justifications for affirmative action as racial redress. *Afr. J. Bus. Ethics* 8, 59–73.

[5] Isidro, H., and Sobral, M. (2015). The effects of women on corporate boards on firm value, financial performance, and ethical and social compliance. *J. Bus. Ethics* 132, 1–19.

[6] Terjesen, S., Aguilera, R., and Lorenz, R. (2015). Legislating a woman's seat on the board: institutional factors driving gender quotas for boards of directors. *J. Bus. Ethics* 128, 233–251.

[7] Koch, R. (2015). Board gender quotas in Germany and the EU: an appropriate way of equalising the participation of women and men? *Deakin Law Rev.* 20, 52–73.

[8] Dale-Olsen, H., Schøne, P., and Verner, M. (2013). Diversity among norwegian boards of directors: Does a quota for women improve firm performance? *Fem. Econ* 19, 110–135.

[9] Brandth, B., and Bjørkhaug, H. (2015). Gender quotas for agricultural boards: changing constructions of gender? *Gend. Work Organ.* 22, 614–628.

[10] Terjesen, S., and Sealy, R. (2016). Board gender quotas: exploring ethical tensions from a multi-theoretical perspective. *Bus. Ethics Q.* 26, 23–65.

[11] Kakabadse, N. K., Figueira, C., Nicolopoulou, K., Hong Yang, J., Kakabadse, A. P., and Özbilgin, M. F. (2015). Gender diversity and board performance: women's experiences and perspectives. *Hum. Resour. Manage.* 54, 265–281.

[12] Lending, C. C., and Vähämaa, E. (2017). European board structure and director expertise: the impact of quotas. *Res. Int. Bus. Finance* 39 (Part A), 486–501.

[13] Holst, E., and Wrohlich, K. (2017). Top decision-making bodies in large companies: gender quota shows initial impact on supervisory boards; executive board remains a male bastion. *DIW Econ. Bull.* 7, 3–15.

[14] Sandberg, S. (2013). *Lean in: Women, Work, and the Will to Lead*, 1st Edn. New York, NY: Alfred A. Knopf.

[15] Wiersema, M., and Mors, M. L. (2016). What board directors really think of gender quotas. *Harv. Bus. Rev. Digit.* 2–6.

[16] Wang, M., and Kelan, E. (2013). The gender quota and female leadership: effects of the Norwegian gender quota on board chairs and CEOs. *J. Bus. Ethics* 117, 449–466.

[17] Rosenblum, D., and Roithmayr, D. (2015). More than a woman: insights into corporate governance after the French sex quota. *Indiana Law Rev.* 48, 889–930.

[18] Klettner, A., Clarke, T., and Boersma, M. (2016). Strategic and regulatory approaches to increasing women in leadership: multilevel targets and mandatory quotas as levels for cultural change. *J. Bus. Ethics* 133, 395–419.

[19] Wyss, B. (2015). Seats for the 51%: beyond the business case for corporate board quotas in Jamaica. *Rev. Black Polit. Econ.* 42, 211–246.

[20] Gephart, R. (2004). Qualitative research and the academy of management journal. *Acad. Manage. J.* 47, 454–462.

[21] de Anca, C., and Gabaldon, P. (2014). The media impact of board member appointments in Spanish-listed companies: a gender perspective. *J. Bus. Ethics* 122, 425–438.

[22] Nekhili, M. and Gatfaoui, H. (2013). Are demographic attributes and firm characteristics drivers of gender diversity? investigating women's positions on French boards of directors. *J. Bus. Ethics* 118, 227–249.

[23] Perrault, E. (2015). Why does board gender diversity matter and how do we get there? The role of shareholder activism in deinstitutionalizing old boys' networks. *J. Bus. Ethics* 128, 149–165.

[24] Rao, K., and Tilt, C. (2016). Board composition and corporate social responsibility: the role of diversity, gender, strategy and decision making. *J. Bus. Ethics* 138, 327–347.

Appendixes

Appendix I: Legislative Measures on Gender Balance in the Corporate Context in Portugal

Legislative measures	Objectives
Resolution of the Council of Ministers No. 49/2007	States that publicly owned companies shall adopt equality plans aiming at promoting equality of opportunities for both genders as well as the conciliation of personal and professional lives.
Resolution of the Council of Ministers No. 70/2008	
Resolution of the Council of Ministers No. 19/2012	Defines the adoption of gender equality measures for management and supervisory boards of state companies, namely through equality plans. Moreover, recommendations are made to private companies listed on the national stock exchange and requests that the state, as a shareholder of private companies, suggests gender equality measures to other shareholders.
Framework law of the independent administrative entities with functions of regulating the economic activity of the private, public, and cooperative sectors, approved by law No. 67/2013	Guarantees gender rotation in the chairs of executive boards and 33% representation of each gender for board members.
Decree-law No. 133/2013	Defines gender diversity in executive and supervisory boards and the promotion of gender equality as part of organizations' social responsibilities.
Decree-law No. 157/2014	States that the selection of board members should promote diversity of qualifications and calls for increases to the number of members of the underrepresented gender (targeted companies should report to the Bank of Portugal what measures they took in this field).
Decree-law No. 159/2014	Establishes that both higher female representation on executive and management boards and greater wage equality are factors for consideration in selecting company applications for EU Cohesion Funds.
Resolution of the Council of Ministers No. 11-A/2015	Instructs members of the government to establish a compromise agreement for gender equality with thirteen publicly traded companies with a goal of 30% participation of the underrepresented gender by the end of 2018.

Source: Commission for Constitutional Affairs, Rights, Liberties and Guarantees (CACDLG) legal advice to draft law No. 52/XIII.

Appendix II: Portugal's 26th Position among EU 28 Countries in the Gender Equality Index*.

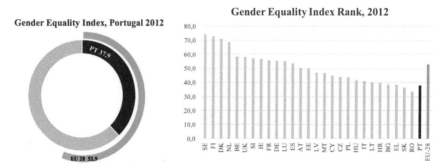

Source: Gender Equality Index, EIGE. Data analysis performed by the author.

*The Gender Equality Index is calculated by the EIGE and assesses gender equality in the EU policy context. Scores from 0 to 100 are available for years 2005, 2010, and 2012. This index includes the following domains: work, money, knowledge, time, power, health, intersecting inequalities, and violence.

5

Reconstructing CSR
in the Construction Industry

**Kwasi Dartey-Baah[1], Kwesi Amponsah-Tawiah[1]
and Yaw A. Debrah[2]**

[1]Department of Organization and Human Resource Management,
University of Ghana Business School, Accra, Ghana
[2]School of Management, Swansea University, Swansea, UK

Abstract

Corporate Social Responsibility (CSR) agenda in Sub-Saharan Africa has
evolved from mere philanthropy to becoming a business tool for gaining
competiveness, with organizations using it as a medium for mitigating the
negative effects of their business operations on society and the natural
environment. While the appreciation of the concept has gained credence
among many industries in Ghana, little is known about its conceptualization,
direction, and drivers in the Construction industry, even though the industry
is noted to be heavily involved in the society. This chapter provides empirical
evidence on the conceptualization, direction, and drivers of CSR in the
Ghanaian construction industry by providing accounts of employees from
over 30 construction firms with operations across the length and breadth of
the country.

5.1 Introduction

In this era of globalization, corporate social responsibility (CSR) is increasingly gaining visibility as it has become expectant of businesses to respond
to emerging social and environmental issues [1, 2]. This is attributable to the
fact that in recent times, a wide array of issues of global concerns has come

181

to the limelight; these include rapid growth in natural resources consumption, the global economic recession, the exploitation of child labor, and the likes [3]. Thus, CSR has become a strategic tool to achieve competitive edge in products and services. In effect, organizations are beginning to appreciate and practice CSR – a construct that lays emphasis on the obligation of organizations to incorporate both social and environmental parameters into their modus operandi and long-term development policies [4, 5].

The burgeoning nature of CSR in this millennium has given rise to a surfeit of definitions for the concept in literature; numerous authors tend to define it based on what they think should be the underlying theme of the concept. The World Business Council for Sustainable Development (WBCSD) in 1999 defined CSR to be the "continuing commitment by business to behave ethically and contribute to economic development while improving the quality of life of the workforce and their families as well as the local community and society at large" [6]. Pearce and Doh [7] contend that CSR encompasses the actions of organizations, which are beyond the expectations of the law and the direct interests of shareholders, aimed at offering some benefits to society. From these, there are some divergent commonalities in the definitions which indicate that socially responsible firms are those that place premium on ethics, respect for stakeholder interests, sustainable development, and respect for the law and human rights. Generally, it is consented that socially responsible organizations enjoy such benefits as increased profitability, improved public image, sustainable competitive advantage, organizational effectiveness, and efficiency, inter alia.

Corporate social responsibility has become a household term among firms across all sectors. However, it is drawing increasing attention in the construction industry because the industry is perceived in the public domain as irresponsible due to its adverse social effects such as the emission of high levels of carbon, resource exploitation and environmental disturbances [2, 8]. Similarly, Sharman [9] contends that the impact of the construction industry on society is significant as the built environment created influences both the physical and psychological well-being of the people therein. In the quest to mitigate and somehow overcome these challenges through technological advancement, contractors or managers in the construction industry are encouraged to promote CSR as a complementary measure [10, 11]. This has fueled the call for the development and promotion of CSR in the construction industry. As a result, there have been several campaigns by governments and environmentalists which exert some considerable amount of pressure on construction firms to act socially responsible while they ensure the effectiveness and efficiency of their building and constructing services as well as management of their business [2, 12].

Despite the shortcomings of the construction industry, it contributes significantly to socio-economic development in diverse ways. In general terms, the industry contributes to the Gross Domestic Product (GDP) of economies, creates employment avenues, provides goods and services, serves as a source of governmental revenue, and so on. Focusing on the impact of the construction industry's activities on society, the economy, and the environment, coupled with the significance of the industry in socio-economic development, the industry ought to pay keen attention to CSR than others [13]. Firms within the construction industry are expected to tackle a myriad of CSR issues such as the nature and status of employment, environmental concerns and relationships with communities, ethical business process, and the likes [8]. Nonetheless, this is not the case as Lin et al. [2] posited that the construction industry lags behind in the promotion and implementation of CSR activities. This could be as a result of the difficulty many construction firms face in aligning their social, ethical, and environmental concerns with their activities and stakeholder interactions [14]. Also, due to the fragmented and diverse nature of the construction industry, just a handful of the firms therein have been able to successfully transform their businesses to be socially responsible ones [15].

In order to improve CSR implementation in the construction industry, Othman [16] suggested that innovative and value-added solutions should be explored. This has championed several discussions on the concept in the industry. Loosemore and Lim [17] contend that discussions of social responsibility in the construction industry are fragmented and confrontational since studies in this regard allude to the fact that CSR initiatives in the industry are heavily determined by organizational culture and adjudged a business case instead of viewing it from the perspective of the beneficiaries whom it's meant for. In furtherance, Loosemore and Lim [17] opined that in the construction industry, CSR is largely uncategorized and highly fragmented covering a wide array of diverse issues. Some of the areas in construction that CSR have covered include community engagement [18], environment and sustainability [19], development indicators [20], stakeholder management [21], and human resource management [22], inter alia.

It is evident that CSR research in the construction industry is becoming increasingly popular and moving forward on numerous fronts; however, Lin et al. [2] posited that this still remains at the initial stages. These previous researches on CSR and the construction industry have promoted the understanding of the concept and its linkage with the construction industry in developed countries. However, apropos of CSR, researches in the construction industry in developing countries (such as Ghana) have not been extensively explored. Again, Loosemore and Phua [23] indicated that the growing literature on CSR in the construction industry is problematic as it is simply pitching CSR as a wholesale solution for all firms without truly understanding the specific business environment in which it was applied. It is therefore imperative to study CSR in the construction industry in Ghana (as the country is undertaking significant infrastructural projects to promote its development) to ensure better understanding of the phenomenon, the factors affecting it, and how appropriately CSR can be applied in this context to ensure sustainable development through sustainable building and construction. In view of this, the chapter seeks to fill the gaps by exploring CSR in the construction industry in Ghana and investigating: (1) the knowledge and understanding of CSR among construction firms in Ghana, (2) the CSR direction of construction firms in Ghana, (3) the drivers of CSR in the Ghanaian construction industry, and (4) how the professions in the construction industry influence the conceptualization and direction of CSR in Ghana.

5.2 Theoretical Underpinnings

5.2.1 Corporate Social Responsibility (CSR)

The term "CSR" came into common use in the late 20th century, around 1960s and early 1970s, after the formation of many multinational corporations. According to Post et al. [24] cited in Enimil et al. [25], the idea of CSR appeared around the turn of the 20th century in the United States. Corporations at that time came under serious scrutiny for being too big, too powerful, and guilty of anti-social and anti-competitive practices. Thus, critics tried to limit corporate power through anti-trust laws, banking regulations, and consumer protection laws. Consequently, a few visionary business executives advised corporations to use some of their power and influence voluntarily for broad social practices rather than for profits alone. One of the wealthiest business leaders like steelmaker Andrew Carnegie is a good example because he took up the challenge and became a great philanthropist who gave much of his wealth to educational and charitable institutions [25]. Others like automaker Henry Ford, developed paternalistic programs such as increased salaries, paying social security taxes, provided conducive environment to work in, support recreational and health needs of their employees. This paradigm shift was a result of the belief by organizations that they have a responsibility to society that went beyond or worked in parallel with their efforts to make profits (Ibid).

The concept of CSR has received diverse definitions by different researchers and this has been identified as a significant problem because it breeds different perspectives of CSR resulting in unproductive engagements [27]. Despite the numerous definitions of CSR, this paper adheres to the definition by the European Union Commission as cited in Ocran [28], which states that CSR is a concept that allows businesses to fuse social and environmental concerns in their business operations and their interaction with stakeholders on voluntary basis. CSR has generated much attention in academia over the years and this is evident in the extensive literature on the concept. For instance, the works of Amponsah-Tawiah and Dartey-Baah [27], Anku-Tsede and Deffor [29], Carroll [30, 31], Carroll and Shabana [32], Moura-Leite and Padgett [33], and Ofori and Hinson [6], inter alia, have made provision for a comprehensive historical development of the concept as well as bibliography of the core academic literature both in the local and global context. In addition, studies have focused on other aspects of CSR such as the significance of the concept (see [34]) and others.

5.2.2 History and Nature of Corporate Social Responsibility in Ghana

Amponsah-Tawiah and Dartey-Baah [35] point out that like many developing countries, the concept and practice of CSR in Ghana was and remains misunderstood. In that, many organizations make donations or give out tokens in the name of CSR either without understanding what it means or merely as a public relation gimmick [25]. In Ghana, due to the practice of a command economy under the socialist orientation of the first republic led by Dr. Kwame Nkrumah, where the responsibility for the production and distribution of good and services was in the hands of State Owned Enterprises, an impression was given that the state through its agencies could address societal problems [35]. However, because of globalization and an ever-growing population with its resultant demand on scarce government resources for goods and services, private organizations – both indigenous and international – have in recent years been allowed to take over a large chunk of the responsibility of providing goods and services.

Corporate Social Responsibility in Ghana to a very large extent is spearheaded by large multi-national companies operating in the various sectors of the economy [6]. This is mainly due to the multi-faceted nature of Ghana's developmental challenges (high poverty levels, low per capita income, weak currency, low productivity, highly agrarian economy, etc.), which makes it close to impossible for indigenous organizations (mostly Small Medium scale Enterprises), most of who are engaged in the retail and production of primary commodities, to undertake CSR [35]. However, recent rankings of the Ghana Club 100 companies have shown an increase in the number of indigenous firms that are engaging in the business of CSR [36].

The concept of CSR has general been understood in the Ghanaian context as philanthropy to the neglect of the ethical, legal, and economic responsibility as identified by Carroll [32]. Many organizations have jumped on the bandwagon of dolling out cash presents, food stuffs, painting of buildings, provision of school stationery, and so on as the pinnacle of their organization's CSR drive. The relatively poor understanding and commitment of CSR has led to what Rockson [37] has described as "philanthropic CSR" or "the Santa Claus way." This he argues is unsustainable and leads to more complex problems for organizations and their beneficiary communities. He postulates that philanthropic CSR as seen in Ghana has led to the emergence of a cadre of receiving communities who do not even have the ability to take over the facilities set up for them by organizations.

5.2.3 Factors that Drive CSR in Ghana

Generally, the factors that influence CSR practice in Ghana can be categorized under two headings: internal factors and external drivers. Internally, socio-economic factors like high poverty levels in Ghana have driven many organizations to direct their CSR activities toward poverty alleviation. Ofori [38] and Enimil et al. [25] posit that the high level of poverty especially in rural Ghana has greatly influence the philanthropic nature of most organizations CSR activities. Anku-Tsede and Deffor [29] also allude that political, social, and economic developments in Ghana over the past two decades, mainly characterized by liberalization, privatization, and democratization, have led to a paradigm shift in the role of organizations toward playing more significant roles in social and environmental problems of their host communities.

Again, the growth of Non-Governmental Organizations (NGOs), community action groups have led to increased responsibility of organizations toward their larger community. Ocran [28] and Ofori [38] have also pushed the argument that CSR in Ghana has to a large extent in recent times being aimed at plugging gaps left by non-performing government and has being largely targeted toward the provision of social amenities that successive governments hitherto have failed to provide in many communities. Externally, issues of International standardization of corporate activities, Stakeholder activisms, and investor demands have served as valuable determinates of CSR practice in Ghana [39].

5.2.4 Sectorial Analysis of CSR Activities in Ghana

Corporate social responsibility activities in Ghana are mostly spearheaded by large-scale multi-national companies doing business mainly in the Extractive, Banking, Telecommunications, and Manufacturing sectors of the economy [27]. The mining industry which boast of organizations like Anglo-gold Ashanti, Cosmos Oil, Goldfields, Tullow Oil, and many more who through their activities contribute toward the development both positively and negatively. In the Banking and Finance industry, the major players include Ecobank, Agriculture Development Bank (ADB), Stanbic Bank, Unibank, HFC, National Investment Bank, SG-SSB, Star Assurance, Vanguard Assurance, etc. The Telecommunication industry boasts of global giants MTN, Vodafone, Airtel, Globacom (GLO), and local firms like Tigo and Expresso. The manufacturing is dominated by Coca-Cola, Fan Milk Ghana, Kasapreko, Interpact, Chocho industries, Nestle Ghana, Cadbury Ghana, Unilever Ghana, Japan Motors Ghana, Toyota Ghana, etc. Due to the role religion and sects play in the social lives of Ghanaians, there are some religious bodies that engage in a lot of CSR activities as well. They include the Catholic Church, Assemblies of God, the Anglican Church, Methodist Church, Pentecost Church, etc. Below is a table showing organizations by sector and their major area(s) of CSR concentration.

From Table 5.1, it is evident that majority of organizations having documented evidence of CSR activities in Ghana are multinational companies. It is also evident that while the mining and telecommunication sectors are more oriented toward social economic development activities such as the provision of social amenities in their areas of operation, religious bodies are more aligned toward the provision of education, charity and relief services to a broader group of the Ghanaian population without any geographical limitation. This is mainly due to the fact that religious bodies especially churches are spread all over the country.

Again, most manufacturing companies in Ghana are pursuing technology development as their CSR probably because they rely heavily on technology in their production process. However, like most organizations in the other

Table 5.1   Selected companies and their corporate social responsibility (CSR) distribution

Sector	Companies	Classification	CSR Area
Mining	AngloGold	Multinational	Health
	Goldfields	Local	Economic development
	Sahara Group	Multinational	Health
	Tullow Oil	Multinational	Economic development
	Goil	Local	Charity
Telecommunication	MTN	Multinational	Health, education, and
	Vodafone	Multinational	economic development
	Tigo	Local	Social amenities, health, and education
	Airtel	International	Charity, social amenities, and health
	Expresso	Local	Health, charity and relief, and social amenities
			Charity
Banking and finance	Ecobank	Multinational	Education and health
	Agricultural Development Bank	Local	Education, health, and staff development
	Access Bank	Multinational	Education, charity, and relief services
	Fidelity	Multinational	Social amenities, education, child development, and health
	Barclays Bank	Multinational	Education, child and youth development, and entrepreneurship
	Ghana Commercial Bank (GCB)	Local	Health and education
	Star Assurance	Local	Education
Manufacturing	Coca-Cola	Multinational	Education
	Chocho Ind.	Local	Charity
	Interplast	Multinational	Technology
	Kasapreko	Multinational	Health and relief service
	Nestle	Multinational	Technology
	Voltic	Local	Education and charity
	Unilever	Multinational	Health and education
Religious bodies	Anglican Church	Multinational	Education, charity, and health
	Catholic Church	Multinational	Education, health, and charity
	Assemblies of God	Multinational	Health
	Pentecost	Multinational	Education, charity, and health
	Methodist	Multinational	Education, health, relief
	Presbyterian	Multinational	services, job creation, etc.
	Ahmadiyya	Local	Health and education
			Education and charity

Source: Authors, 2017 (Adapted from Amponsah-Tawiah and Dartey-Baah, 2016).

sectors, especially the local ones, organizations in the manufacturing sector have charity as a key component of their CSR initiatives. Even though it is evident from the table that most multinational companies and a few number of indigenous firms are beginning to have a strategic outlook to their CSR, there is sufficient evidence from Table 5.1 that gives credence to earlier assertions by numerous authors [6, 28, 29, 35, 36, 38] that CSR in Ghana is to a large extent philanthropic in nature. Again from Table 5.1, it is evident that CSR activities in the construction industry in Ghana have not been highlighted in most studies. This offers credence to the fact that the practice of CSR lags behind in the industry, thus the importance of this study.

It can be noted from Table 5.1 that CSR activities of construction firms which have been described by Pitt et al. [40] as the "locomotive" of all the other industries have not been presented. Even though the review of literature on CSR shows that extensive work has been done across various sectors, especially with regard to the CSR activities of firms in Ghana, there is a lack of comprehensive empirical research on the nature and scope of CSR in the construction industry. Particularly, literature on the direction and drivers of CSR in the construction in Ghana remains untouched. This study therefore serves as a groundbreaking study that explores CSR within the construction industry.

5.2.5 Institutional and Regulatory Framework of CSR in Ghana

Over the years, CSR in Ghana has being expressing a paradigm shift from just being a voluntary practice to a more mandatory practice influenced by public policy and law [29]. Even though Ghana has no direct laws regulating CSR, the importance of the presence of other laws such as the Environmental Protection Agency Act, 1994 (Act 490), the Labour Act, 2003 (Act 651), and a host of other legislation in institutionalizing and regulating the actions of organizations in relation to their immediate and remote society cannot be over-emphasized.

Currently, there is no existing comprehensive legal framework or legislation to guide the practice of corporate responsibility in Ghana, though there are relevant laws ensuring that organizations conduct their businesses responsibly. However, Atuguba and Dowuona-Hammond [36] and Anku-Tsede and Deffor [29] contend that the lack of coherence in these laws has inadvertently created problems of implementation for regulatory bodies, which in the long run result in CSR activities of organizations still remaining voluntary in practice. Again, until 2006 when the Ghana Business Code

(GBHC) was launched with the aim of deepening the practice of CSR business operation, there was no set of norms and ethics to guide the conduct of business responsibly with regard to environmental protection and counter corruption in business [35].

Furthermore, organizations are not obliged by law to sign up to the GBHC. It is not surprising to note that in spite of the many indigenous and multinational firms registered under the Ghana National Chamber of Commerce and Industry (GNCCI) and the Association of Ghana Industries (AGI), less than 60 organizations had signed up to the GBHC by the first quarter of 2011 [35]. The call has thus been for government to pass a legislation on CSR which provides a guide to business operation, provides incentives and punishment for organizations with their operation in Ghana.

In 2005, NEPAD's African Peer Review Mechanism recommended the clarification of the concept of CSR in Ghana in an effort to promote it among all stakeholders involved. Owing to this, Government of Ghana in partnership with the German Development Organization (GIZ-Ghana) and the Center for Cooperation with the Private Sector, South Africa, have concluded the process of developing a national policy on CSR. This policy that will serve as a comprehensive national policy to guide business operations in the country is set to be launched later this year, paving way for a more targeted, coordinated, and legal underpinning for the practice of CSR in Ghana.

5.3 Methodology

This chapter is based on a study of CSR in the construction industry in Ghana. The study adopted an in-depth exploratory approach to operationalize the research questions. The design was therefore cross-sectional in nature as the study sought to gain a snapshot of the study population. The population consisted of employees in construction firms operating in Ghana; 100 questionnaires were distributed to the respondents; however, 71 valid questionnaires were retrieved, and thus, a response rate of 71%.

For the purpose of the study, a detailed open-ended questionnaire was developed. The instrument was divided into three sections. The first section collected information on the demographics of respondents (such as educational qualification, profession, name of firm, position, experience, etc.). The second and third parts of the instrument allowed respondents to describe in their own words their understanding of CSR, the direction of their firm's CSR, and what they felt about the influence their profession had on their conceptualization of CSR and their firm's CSR direction. The instrument was initially given to HR managers of five top construction firms in Ghana to validate in accordance with the study's aims and objectives. This enabled the instrument to be checked for content and face validity, and reliability prior to the actual data collection.

Given the exploratory nature of the research and to answer the main research questions, an in-depth content analysis and clustering of findings from each question and section was implemented. This was to help the researchers map out salient nuances with respect to the study variables. A combination of descriptive accounts given by respondents sampled as well as summary statistics in terms of frequencies and percentages was used.

5.4 Results and Discussion

In light of the fact that the data instrument used for this paper was divided into three sections (Respondents' demographics, CSR perception and direction of firm, and Profession's influence on CSR practice), our findings are presented with reference to the above-mentioned sections. The findings are presented using a combination of summary statistics (frequencies and percentages) and hand-merged themes derived through careful analysis of the answers given by respondents.

5.4.1 Respondent Demographics (Section 1 of the Instrument)

The following is a summary of the demographic characteristics of respondents sampled:

- Out of a total of 71 valid questionnaires analyzed, 43 (61%) respondents worked in firms that practiced CSR as part of their operations.
- These respondents represented 36 firms operating in Ghana. Of this number, only six did not fit the classification of construction or engineering companies, even though the respondents sampled from these firms performed tasks tied to either construction or engineering.
- All respondents worked in either technical or non-technical positions for their respective firms, with 61% working in technical positions (Civil, Mechanical, Electrical, Geodetic Engineers, Project Managers, Quantity surveyors, and Procurement) while the remaining 39% worked in non-technical positions (Finance and Accounting, Sales and Marketing, Human Resource Management, Supervisors, Administration, Fleet management, etc.).

- More than 69% of respondents had attained at least a first degree, while 14% have post-graduate degrees. The remaining 16% had at least a professional certificate from an accredited institution in Ghana.
- With respect to work experience, 34 (79%) respondents had spent between 1 and 5 years in their current profession, while 7 (16%) respondents and 2 respondents (5%) had worked for 5–10 years and over 10 years in their current profession, respectively.

5.4.2 Perspectives on CSR among Construction Workers (Section 2 of the Instrument)

Findings under this section summarize responses given by sampled construction workers on their:

   i. Knowledge and conceptualization of CSR
  ii. Corporate social responsibility direction of their firm
 iii. Drivers of CSR in their firm.

5.4.2.1 Knowledge and conceptualization of CSR

The data analysis focuses first on examining respondents' views with regard to their knowledge and conceptualization of CSR. The responses are further grouped under four CSR dimensions as proposed by Carroll [32] and Maignan et al. [41], with the aim of identifying the most dominant conceptualization of CSR among respondents.

In sum, respondents showed a fairly good level of knowledge and understanding of the concept of CSR. An HR manager sampled explained that:

> *"CSR promotes the concept that a business has an obligation to do more than just generate profits for its owners. It could be embracing a charity or following environmentally sound practices in its business activities."*

A few other respondents described it as a legal and ethical business responsibility – a platform through which their firm engage with communities within which they operate and the Ghanaian society at large. The response of a geodetic engineer captures CSR as:

> *"... a way of corporations' self-regulating themselves by integrating compliance and ethical standards, national or international norms, and the spirit of law in their operations. In simpler terms,*

it refers to how conscious companies are and about how their operations impact not only on their profits but also the society and environment. This makes CSR a means through which organizations give back to the society within which they operate."

Most respondents, however, viewed it as cost firms had to shoulder, though voluntary and sparingly, to ensure that they meet the needs of the communities within which they operate. This they noted is done through the provision of social infrastructure (i.e., school buildings, clinics, pipe-borne water, etc.) and donations to support community programs (scholarship schemes, festivals, etc.). A respondent (Chief Executive Officer of a sampled firm) uses the term "appreciation" to describe his understanding of the concept.

"Allocating part of our limited resources occasionally toward making meaningful contribution to the community of our operations as well as other laudable initiatives aimed at improving the lives and showing 'appreciation' to the chiefs and people in the community."

Even though the results show a fairly good understanding of the concept of CSR among workers in the construction industry, the responses indicate a shallow understanding of the concept. As argued by scholars like Ofori [25, 38], Amponsah-Tawiah and Dartey-Baah [27], many organizations in Ghana view CSR as an external element separate from their core organizational processes and practices. These scholars noted that this understanding of CSR is mainly due to political, socio-economic, and cultural environment within which Ghanaian firms operate. As reflected in the responses of the workers sampled, the construction industry like other industries in Ghana views CSR as a cost borne by organizations in an attempt to deal with social, political, and environmental issues sometimes created by virtue of their operations and not as a tool for gaining shared value for both the firm and society. As Enimil et al. [25] put it, CSR in Ghana is steeply grounded in the culture of successful people giving back to the community as a means of showing appreciation. This understanding of CSR as an external practice is probably a reason why many Ghanaian firms do not view CSR as internal business process. Particular reference is made to issue of Occupational Health and Safety concerns that continue to plague the industry. This is against the backdrop that the construction industry is known to be a highly hazardous industry, and thus issues of Occupational Health and Safety should be high on the agenda of firms and not copiously missing from the industry's modus operanda as seen in the particular case of Ghana's construction industry. This is important because Occupational Health and Safety (OHS) concerns have the tendency of impacting not only the bottom line of the firm but also the image and reputation of the firm in the public as an organization that prioritizes the safety of its workforce.

As indicated earlier, the responses given by respondents were further categorized using Carroll [43] and Maignan et al. [41] conceptualization of CSR. As shown in Table 5.2, it is evident that CSR is conceptualized as a philanthropic or discretionary activity by majority of workers in the construction industry in Ghana with as many as 20 respondents providing definitions that reflect the view that CSR is a benevolent gesture to communities within which they operate. This is followed by ethical (17), economic, and legal with three responses each.

Clearly, CSR in the construction industry continues to be in its embryonic stage in terms of its conceptualization, with many players in the industry conceptualizing it only within the context of philanthropy and ethics. The construction industry in Ghana has gained a generally poor ethical reputation,

Table 5.2   Distribution of CSR conceptualization per Carroll's (1979) Pyramid

CSR Dimension	Frequency	Respondents	Sample Definition
Economic	3	14, 28, 41	My own understanding of CSR is about how companies engage with their stakeholders to manage the business processes to produce an achievable economic outcome.
Legal	3	10, 27, 38	Corporate social responsibility is a way of corporations self-regulating themselves by integrating compliance and ethical standards, national or international norms and the spirit of the law in their operations...
Ethical	17	2, 8, 9, 13, 21, 22, 25, 26, 29, 31, 34, 35, 37, 39, 40, 42, 43	It is a means by which corporate organizations go beyond their legal obligations to preserve and protect the society and environment within which they operate.
Discretionary	20	1, 3, 4, 5, 6, 7, 11, 12, 15, 16, 17, 18, 19, 20, 23, 24, 30, 32, 33, 36	It is the effort of some corporate companies to improve the society by helping in many ways (donating money, building facilities, and implementing programs).

Source: Author's construct, 2017.

being widely regarded in the public as an industry inundated with corrupt practices, health and safety failures, and bad environmental practices [44]. This probably is the reason why construction firms in Ghana place high premium on ethical CSR. Again, the results indicate that firms in Ghana do not see the legal domain as a reason for CSR engagement. However, while the ethical orientation remains common, increased attention on economic CSR could increase the attention accorded to legal CSR. Results also corroborate earlier findings of Yeboah [45] and Rockson [37] who reported that CSR in the Ghanaian context is seen as philanthropic to the neglect of the ethical, legal, and economic responsibility with organizations jumping on "the bandwagon of dolling out cash presents, food stuff, painting of buildings, provision of school stationery" and viewing that as the pinnacle of their social responsibility. This reveals that while the companies point out mostly to ethical reasons in their CSR understanding, the philanthropic orientation remains the primary basis of CSR adoption among construction firms in Ghana.

5.4.2.2 CSR direction of construction firms in Ghana

It was discovered that though the extent of the CSR activities undertaken by construction firms vary based on the nature of their activities and firm size, their CSR directions remained similar. As indicated by respondents, the CSR direction of construction firms in Ghana is focused mainly on activities in the areas of education and skill development, healthcare, environmental protection and sanitation, and socio-economic development, among others. This is probably because the nature of the activities performed by construction companies leaves a lot of people especially in poor communities deprived of their customary livelihoods mainly because of the pollution of river bodies and the environment as well as the loss of arable lands that support their basic livelihoods. Again, the scope of their CSR activities, just like in the telecommunication and mining industries in Ghana, is limited not only to their immediate communities but also to the larger Ghanaian society.

As expected, majority of respondents indicated that their firm's CSR activities were directed toward the support of education and skills development, with substantial investments either in the provision of building materials for the construction of schools, or building the schools themselves, provision of scholarships for needy students, donation of books and other educational materials to schools within their operational areas, as well as the sponsorship of staff to upgrade their educational qualifications.

"Our CSR focus is mainly geared toward the education sector. This is because many of the children we encounter in the communities are uneducated, sometimes due to poverty or the unavailability of schools in the community. We try our possible best to support the sector to make our communities better. The policy is in different forms; to sponsor some of the workers who have the ambition to further their education and provide some educational materials and facilities needed by schools."

"Our policy is directed toward enriching poverty stricken individuals with computer skills deemed imperative. Again, my organization seeks to equip women with skills such as sewing to enable them fend for themselves and their families."

With regard to environmental protection and sanitation, respondents indicated that their firms found it ethically important to ensure that their activities do not harm the environment or cause harm to the lives of inhabitants. Consequently, some respondents indicated that their CSR policies were based on strict ethical requirements and best practices governing the construction sector, which includes ensuring that construction projects, such as roads, bridges, etc., provide value for money and the right quantity of materials is used to forestall accidents that may lead to the loss of human lives and property. Again, attention is given to ensure that construction projects have minimal effects on the natural environment, and in case not, corrective measures are taken to address any harm caused. Two civil engineers working on projects in the KEEA district in the Central region of Ghana asserted that:

"Being a construction company, we have a significant impact on the communities we build. The way we design and build houses and other infrastructure has social impact requirements which ensure minimal impact on the environment as well as ensures that projects span through decades."

"As a convention, my firm has tree planting attached to all our project plans. We do this knowing the effect of our construction projects on the physical environment which always involves cutting done some trees and destabilizing the eco-system. We therefore do not charge our clients for the tree planting activities."

On the health front, as seen in the response below, some of the respondents revealed that their firms' tend to focus on the construction of wards and rehabilitation of dilapidated hospital structures in communities, drilling bore-holes, constructing Kumasi Ventilated-Improved Pit (KVIP) toilets for some communities, making donations to hospitals, as well as footing medical bills of patients.

> *"My organization seeks to help upgrade the well-being of people through health. We focus on building community hospitals organizing free health screening, and contributing to foundations in the country. For example, we have been making regular donation to the Joyce Tamakloe Memorial Cancer Foundation since November 2007."*

In terms of socio-economic investments, a few respondents reported that their firms embarked on some form of social investments activities as part of their CSR. This includes but not limited to building of community centers and storage facilities to forestall perianal loss of harvests.

As reported in an earlier study by Amponsah-Tawiah and Dartey-Baah [27], the multi-faceted nature of the problems in Ghana, which include but not limited to "low per capita income, high population rate, weak currency, capital flight, low productivity, low savings, etc.," (p. 9) makes it imperative for firms to direct their CSR initiatives toward solving such problems. It is thus not surprising that the CSR of firms in the construction industry are similar to that of firms in the extractive, telecommunication, manufacturing, services, and banking sectors. This follows from the point that Kuada and Hinson [46] had reported in an earlier study that CSR in Ghana is hugely influenced by socio-economic, political factors and development gaps. Again, the relative attention given to environmental issues by construction firms ties directly to the nature of their daily operations, which usually involves the moving of earth and the destruction of vegetation, water bodies, and their natural habitat, which ultimately has a dire effect on the livelihood of locals who depend on the environment for their basic livelihood. As Pitt et al. [40] put it, "the built environment affects all human activity." In addition to this, the industry is a major facilitator and contributor to the overall economy. Consequently, the construction sector has major impacts on all three pillars of sustainable development: environment, society, and economy [40, 46].

5.4.2.3 Drivers of CSR in the construction industry

Various organizations have different motivations for engaging in CSR. While some are driven by humanitarian and socio-economic reasons, others are driven by stakeholder pressures, and legal and institutional dictates. A good number of firms are also driven by environmental sustainability factors, whereas others are driven purely by management discretion. The analysis revealed five drivers of CSR in the construction industry. These are

1. Nature of firm's operation
2. Environmental sustainability factors
3. Stakeholder and legal and institutional pressures
4. Humanitarian and Human Rights reasons
5. Management discretion.

5.4.2.4 Nature of firm's operation

Even though the construction industry has been applauded for the positive impact of its operations on infrastructural development and economic growth in Ghana, the industry like the extractive industry has been noted to engage in activities that ultimately destroy vegetation and water bodies, leaving in its wake huge socio-economic, environmental, and other health-related ills. As alluded to by some respondents, the nature of the industry's operation, which includes the clearing of vegetation, sand wining, pollution of rivers and water-bodies, displacement of community settlements, and loss of farmlands, among others, not only destabilizes the physical environment but also affects individuals and communities who depend on them. Firms in this industry have thus taken to investment in CSR activities as a tool for addressing the numerous social and environmental problems attributed to the industry. To put it rightly, a chief site engineer of one of this firms sampled attests to this fact. The response is as follows:

> "...the nature of our work makes it imperative to engage in activities that provide a bit of relief for affected individuals and communities. We understand the harm that community design and construction has on the life of residents, so our ultimate driver is to make the society better place to dwell in."

5.4.2.5 Environmental sustainability factors

The past decade has seen citizens and civil society organizations across the world advocating for corporations to pay greater attention to environmental issues in the course of their operations. Issues of climate change, loss of forestlands, pollution of water bodies, and its resultant adverse effects on aquatic life have necessitated such calls. Environmental sustainability and sanitation factors have thus become very important drivers of CSR for the majority of construction firms in Ghana with over 30% of respondents singling them out as the main driver of their firm's CSR. For many of the respondents sampled, CSR has become an effective means of addressing some of the construction-related environmental issues, especially because their firm's activities are tied directly to the environment. Construction firms in Ghana are thus employing measures that include massive tree planting, landscaping, dredging, and the use of plastic, paper, and other non-degradable substances for construction works. Some respondents revealed that their organizations had taken conscious efforts to manage and reduce the use of hazardous substances in their operations, as well as adopting international best practices for the disposal of construction-related waste. The responses of two respondents who noted environmental sustainability factors as their firm's main CSR driver are captured below:

> *"Our CSR is motivated by the impact of our work on environmental degradation, climate change, pollution, etc., and their effect on the society [environment]. We plant trees and for that matter adopt greening the country to in a way solve these problems."*

> *"For us, our CSR is directed by efforts meant to save energy and natural resources, cutting waste and limiting the use of hazardous substances in our operations. Our goal is to manage and recycle waste like plastic and other non-degradable substances by using them in the construction of roads, gutters, and others."*

5.4.2.6 Stakeholder and legal and institutional pressures

As seen in CSR literature, an organization's stakeholders are not only its shareholders. Thus, stakeholders include all interested parties in the existence and operations of the organization. In this sense, those who are not involved directly in the firm's regular decision-making process and yet are affected by the existence of the organization may be included in this classification. Examples of such groups may include the government through its institutions and regulatory agencies, customers, suppliers, media and pressure groups, etc. The analysis of the respondent's responses indicated that pressures from institutional and regulatory bodies like the Environmental Protection Agency (EPA), media reportage, and community-based CSOs have become a major motivation for the adoption of CSR as a means of meeting balancing the demands put on the organization. For instance, a respondent reported that it is now imperative for construction firms to provide impact assessment reports of construction projects before and after they are done before payments are effected for the projects.

Again current interest of the Ghanaian media in infrastructural projects, which is evident from the regular media reportage on infrastructural projects across the country, has made it difficult for firms in the industry to waver on their ethical and legal responsibility to stakeholders who may have either direct or indirect interest in the project. A project manager encapsulates this simply:

> *"There are a lot of pressures from various stakeholders especially with public projects. Apart from the provision of an impact assessment reports to the EPA regularly, we also have to deal with demands for civil society and other pressure groups like the Association of People with Disability to make sure our projects are friendly to their members. Also any negative media reportage can get your firm to be blacklisted by the government so we are very careful to ensure that the right things are done each step of the way. We therefore engage in CSR to manage these interests and show our stakeholders that we are responsive to their concerns."*

5.4.2.7 Humanitarian and Human Rights reasons

Humanitarian issues remain the number one driver of CSR in Ghana and the analysis of responses of sampled respondents has confirmed that humanitarian reasons, mainly as a result of socio-economic factors, continue to be the main driver of CSR in the construction industry. Majority of respondents (over 45%) indicated that the realities of poverty and hardship they encounter in the communities in which they work are the main motivation for their CSR activities and programs. Some respondents also indicated that these conditions go to the core of the human rights of the individuals as it pertains to issues of access to education, healthcare, portable drinking water, and basic livelihood. Some construction firms as part of their CSR resort to the building of classroom blocks, clinics, provision of seedlings for staples to residents and paying for extension services for farmers, construction of boreholes, etc., as a means of lessening the socio-economic burden of people in their operational communities. The comment below is a response from a Director who works for a construction firm working in Northern Ghana:

> *"The high levels of poverty and illiteracy in the communities we work has steered the organization into undertaking such policies. It is disheartening when we encounter so many people stuck in abject poverty due to the lack of basic social amenities like hospitals, portable water, and schools. We operate under the persuasion that*

if, for instance, illiteracy should be reduced and people are at least given some extent of very good education, be it formal or informal, the level of dependency will reduce while the level of growth in these communities will increase... we also educate farmers and provide them with farming supplements so that they can fend for their families."

While some respondents view it as an ethical responsibility to act on such matters; others indicated that their organizations approached its CSR strictly from philanthropic point view. This is captured in these responses below:

".... Ethics. We deem it ethical to support intelligent but needy children in the community."

"What informs the direction of our CSR is simply to give back to the society because we in the capacity to do so."

5.4.2.8 Management discretion

It is important to note, however, that a few respondents revealed that their organizations did not have any philosophy or motivation underpinning their CSR activities. They attributed this to the fact that CSR activities of their firm (apart from the legally mandatory activities regarding their operation) are done at the whims of management and not necessarily based an institutionalized policy or convention. CSR in such firms is thus characterized by a "once in a long time" donation toward a cause that the director, owner, or manager felt was important. This a respondent alluded is a publicity stunt adopted by the organization to keep itself relevant and control growing public dissatisfaction among sections of residents of their operational communities. This is what one site manager noted:

"I do not think my organization has a particular motivation for engaging in CSR. Once in a while we do some donations to churches and youth groups in the community and organize a 'gala' or make donations to support a program. This is usually when such groups approach management about issues they are dissatisfied about and has no direct bearing on the actual needs of the community."

The results give credence to earlier studies by Anku-Tsede and Deffor [29] and Atuguba and Dowuona-Hammond [36] assertion that CSR in Ghana has been experiencing a paradigm shift from just being a voluntary practice to a more mandatory practice influenced by public policy, external stakeholders, humanitarian and human right issues, and law. For instance, Anku-Tsede and Deffor [29] note that even though Ghana has no direct laws regulating CSR, the importance of the presence of other laws, i.e., Environmental Protection Agency Act, 1994 (Act 490), the Labor Act, 2003 (Act 651), and a host of other legislation in institutionalizing and regulating the actions of organizations in relation to both their internal and external stakeholders has ensured that firms in Ghana, irrespective of the sector, imbibe basic legal and stakeholder concerns in their operations. Indeed, Atuguba and Dowuona-Hammond [36] and Anku-Tsede and Deffor [29] agree that even though there is currently no existing CSR policy and a comprehensive legal framework or legislation to corporate responsibility in Ghana, there are relevant laws ensuring that organizations conduct their businesses responsibly. They, however, lament that the lack of coherence in these laws which has inadvertently created problems of implementation for regulatory bodies, which in the long run result in CSR activities of organizations still remaining voluntary in practice. It is, however, interesting to note that these drivers discovered mainly constitute internal drivers of CSR and give no indication of the external factors that drive CSR in the construction industry.

5.4.3 Profession's Influence on Firms' CSR Practice (Section 3 of the Instrument)

This section presents the analysis and results regarding the level of influence profession has had on the conceptualization of CSR of respondents and the direction of their firms' CSR. The analysis is thus presented as follows:

- The Influence of profession on respondents' conceptualization of CSR
- The Influence of respondents' profession on firms' CSR direction.

5.4.3.1 The influence of profession on respondents' conceptualization of CSR

Majority of respondents believe their current profession has had an influence on their knowledge and conceptualization of CSR with 33 (77%) respondents answering "Yes" when asked if they think their profession has any influence on their understanding of CSR. When asked how? Majority respondents indicated that their profession had given them a better understanding of the effects and importance of their daily activities on the lives of the people in the communities in which they work. A mechanical engineer explained that:

> *"Because I have come to understand that in my line of work, I need to make sure things are done in a proper manner so as to have a positive impact on society."*

Others also indicated that by the nature of the skills they possess by virtue of their professional training, they are able to identify issues within their operational communities, which they can help solve through their work.

> *"Being a marketer in the real estate industry, the core function of my role is problem identification and devising ways to solve them."*

> *"Civil engineering work can be described as the design and execution of construction works to suit the environment. This is because construction activities to some extent have an impact on the environment, so we owe it to the community to take responsibility for the adverse effects of our work. I have therefore come to appreciate the need for my work to solve environmental problems and not to create new ones."*

> *"Engineering profession involves sustainable development. The kind of development we do therefore always takes the environment and society into consideration so as to meet the present needs of society without compromising the ability of future generations of meeting theirs. Also, I usually work in peri-urban or developing communities where there are little or no social amenities. So I experience how important those amenities and donations are to the needy communities. This has increased my level of appreciation of the importance of business intervention in the lives of people."*

For some others, the ethics and standards governing their profession inherently make them inclined to exhibiting socially responsible behavior in the course of their work. An accountant for a construction firm explains the influence of financial ethics on his conceptualization of CSR.

> *"I always seek to ensure that as an accountant, my professional ethics permeates all my activities. For instance, I know it is wrong to authorize payments for materials that do not meet the standard quality criteria since it might have dire repercussions on human lives and property later."*

Respondents who indicated that their profession has no influence on their conceptualization of CSR underlined their responses with two simple reasons. The first is the fact that regardless of one's profession, they deemed it imperative for individuals to be ethical and socially responsible in every aspect of their work, especially the aspects of their work that has the potential to affect other people. Secondly, some respondents revealed that their professional training does not involve CSR and thus they have only been inclined to be conscious about social responsibility because the firm in which they work practice CSR.

5.4.3.2 The influence of respondents' profession on firms' direction of CSR

A greater number of respondents, however, indicated that their profession had no effect on their firm's CSR direction with (68%) of sampled respondents answering "No" when asked whether their current profession has had an influence on the direction of their firm's CSR. Respondents revealed that the CSR policies in their respective firms are documents they inherited and, thus, have no influence on its content or the kind of CSR the firm engages in. Others also indicated that their organization's CSR is normally borne out of the realities encountered in the operational communities and not by any single individual's profession. A few others, however, argued that CSR initiatives of their firm are solely based on the dictates of management and such decisions are made at the boardroom level, and hence their contribution to such is virtually non-existent. These assertions are carried in the responses presented below:

> "Corporate Social responsibility is vast and takes on different forms. However, in the construction firm where I work, the CSR policy is a result of the nature of the realities in the communities of our business operations. My profession certainly has no influence on the direction of CSR."

> "It is the company's executives that decide on which of the social responsibility activities will be fulfilled yearly. Decisions are taking by top-level management. We give inputs sometimes, but not in our professional capacities so the [CSR] direction comes from the authorities."

Few respondents, however, indicated that their profession influenced the CSR directions of their firms. Interestingly, the results indicated most of these respondents work in non-technical positions vis-à-vis the construction industry and thus are normally in-charge of administrative and policy-making and implementation aspects of the work of their firms. A respondent from a Marketing and Sales department of a real estate company sampled explained how her profession has an impact on her firm's CSR direction:

"This is because my department is in charge of designing and implementing the CSR policy and initiative of my firm. So my profession as sales person has a direct impact on the CSR direction of my firm."

A geodetic engineer (technical position) approaching it from an operational point of view, however, explains how his day-to-day duties influence the CSR practice of his firm:

"The direction is in our practices and I think my profession has a major influence on the direction of the CSR activities. This can be seen from the point that I am responsible for advising management on the kind of chemicals to use that would rather protect the environment instead of impacting it negatively."

The results show how disconnected the technical workers in the construction industry are from CSR policy making and probably why issues of Occupational Health and Safety did not come up in the analysis of the CSR direction of their firms. An earlier study on CSR practices in Ghana by Ofori and Hinson [6] reported similar findings. Their study showed that directors of marketing, finance, and human resources, as well as the managing directors and their deputies, of firms in Ghana are those that determine the CSR direction of their firms. This probably explains why employees in technical positions within the construction firm allude that their profession had no direct impact on the CSR direction of their respective firms.

5.5 Implication and Conclusion

The results found though tentative are still promising. Respondents across the construction industry did report that there was low awareness of CSR in the construction industry. On the other hand, although there are differences in approaches all firms take with regard to their CSR types and dimension, ethical and the social orientations have become prominent for CSR initiatives in the industry. Also, the study found that only very few firms in the construction industry are thoroughly implementing CSR. There is therefore still much room for improvement in CSR awareness for the construction industry. For instance, even though most of the respondents sampled showed high awareness of discretionary and ethical CSR as well as an enthusiasm about environmental protection and socio-economic issues, the analyses also show that the profession of employees in the industry has an impact on their knowledge and conceptualization of CSR even though their profession (especially that of technical workers) does not affect their firm's CSR direction.

Nevertheless, within these limitations, the results indicate that CSR in the construction sector is largely informal, unstrategic, unsophisticated, narrowly focused, and still at the embryonic stage. Although all firms see the benefit of CSR, the link to strategy is unclear. More research is needed to investigate any links (both real and perceived) between CSR and business performance in the construction industry in Ghana and to explore the barriers to attaining competitiveness that is central to achieving CSR objectives in such a fragmented industry. The results also show that although CSR is on the business agenda of construction firms, there is little evidence that firms are willing to put it at the center of their business and that there is a fear that if they did so, they would lose rather than gain competitive advantage in the market. CSR initiatives appear to be only externally focused and reflect a traditional philosophy that largely ignores the mutuality of interests between both internal and external stakeholders in the success of organizations. Also, there is little sense of strategic focus in our results and the potential social capital that could be gleaned from building better strategic relationships with communities appears to be ignored.

These results have several implications for construction management research and practice. First, in moving construction CSR research forward, we suggest that the project-based nature of construction work with no emphasis on issues of occupational health and safety, internal stakeholder issues, and the unstrategic nature of the industry's CSR activities guide future CSR research. Current models of CSR such as Carroll's (1991) pyramid, while useful, remain limited due to its static nature and its inability to reflect how firms operating in the construction industry align their CSR strategies to fit the social, environmental, and economic profiles of local communities in which they operate in order to create shared value.

For the purpose of practice, our results indicate that firms in the construction industry could benefit significantly from following some core principles of good CSR practice. This includes choosing CSR causes that align with a firm's core business values and goals, inculcating issues of occupational health and safety and CSR reporting. Not only will this potentially maximize returns on the firm's CSR investments but also ensure that firms will be more transparent and legal in their dealings and treated with less skepticism as well as taken more seriously by stakeholders. Focus should be on the use of limited resources effectively and development of a strong partnership between community, firms, employees, and other external stakeholders. This comes on the footing that CSR within the construction industry is largely conceived as an external element, thus though, for instance, the construction industry is known to be a highly hazardous industry, issues of Occupational Health and Safety remain copiously missing and undervalued. Again, internal stakeholder management issues like employee voice, employee involvement, and participation have not been given critical attention in the industry. Hence, there is a need for firms in the industry to encourage the participation of technical workers in CSR policy formulation and implementation because not only they deeply involved in the work within the local communities, the nature of their work exposes them to many health and safety-related risks and hazards, and thus they have a better appreciation of the issues that need the firm's attention. This is important because it allows the organization to develop CSR policies that focus on issues, which are of concern to the community and internal stakeholders, rather than imposing the firm's priorities on them. Finally, it is important to choose sustainable causes because consistency of focus and dedication to a cause maximizes the chance of building strong and meaningful partnerships with communities, and foster credibility with stakeholders while ensuring the long-term sustainability of CSR initiatives.

References

[1] Schultz, F., Castello, I., and Morsing, M. (2013). The construction of corporate social responsibility in network societies: a communication view. *J. Bus. Ethics* 115, 681–692. doi: 10.1007/s10551-013-1826-8

[2] Lin, X., Ho, C. M. F., and Shen, G. Q. P. (2017). Research on corporate social responsibility in the construction context: a critical review and future directions. *Int. J. Constr. Manage.* doi: 10.1080/15623599.2017.1333398

[3] Lu, W., Wang, W., and Lee, H. (2013). The relationship between corporate social responsibility and corporate performance: evidence from the US semiconductor industry. *Int. J. Prod. Res.* 51, 5683–5695.

[4] Persais, E. (2002). Ecology as a strategic asset: A validation of the resources approach by the method PLS [Ecology as a strategic asset:

A resource validation approach by PLS method]. *Rev. Finance Control Strategy* 5, 195–230.

[5] Martínez, P., Pérez, A., and del Bosque, I. R. (2013). Measuring Corporate Social Responsibility in tourism: development and validation of an efficient measurement scale in the hospitality industry. *J. Travel Tour. Market.* 30, 365–385.

[6] Ofori, D. F., and Hinson, R. E. (2007). Corporate social responsibility (CSR) perspectives of leading firms in Ghana. *Corp. Gov.* 7, 178–193.

[7] Pearce, J. A., and Doh, J. P. (2005), The high impact of collaborative social initiatives. *MIT Sloan Manag. Rev.* 46, 30–38.

[8] Barthorpe, S. (2010), Implementing corporate social responsibility in the UK construction industry, *Prop. Manage.* 28, 4–17.

[9] Sharman, J. (2016). *Social responsibility and the construction industry.* Available at: https://www.thenbs.com/knowledge/social-responsibility-and-the-construction-industry [accessesd July 11, 2016].

[10] Shen, L. Y., Tam, V. W. Y., Tam, L., and Ji, Y. B. (2010). Project feasibility study: the key to successful implementation of sustainable and socially responsible construction management practice. *J. Clean. Prod.* 18, 254–259.

[11] Jiang, W., and Wong, J. K. W. (2015). Key activity areas of corporate social responsibility (CSR) in the construction industry: a study of China. *J. Clean. Prod.* 113, 850–860.

[12] Petrovic-Lazarevic, S. (2008). The development of corporate social responsibility in the Australian construction industry. *Constr. Manage. Econ.* 26, 93–101.

[13] Murray, M. and Dainty, A. (2008). *Corporate Social Responsibility in Construction Industry.* Abingdon: Taylor & Francis.

[14] Duman, D. U., Giritli, H., and McDermott, P. (2016). Corporate social responsibility in construction industry. A comparative study between UK and Turkey. *Built Environ. Project Asset Manage.* 6, 218–231.

[15] Myers, D. (2005). A review of construction companies' attitudes to sustainability. *Constr. Manage. Econ.* 23, 781–785. doi: 10.1080/01446190500184360

[16] Othman, A. A. E. (2009). Corporate social responsibility of architectural design firms towards a sustainable built environment in South Africa. *Architect. Eng. Des. Manage.* 5, 36–45. doi: 10.3763/aedm.2009.0904

[17] Loosemore, M., and Lim, B. T. H. (2017). How socially responsible is construction business in Australia and New Zealand? *Proc. Eng.* 180, 531–540.

[18] Teo, M. M. M. (2008). *An Investigation of Community-Based Protest Movement Continuity against Construction Projects.* Unpublished Ph.D. thesis, FBE, UNSW, Sydney.

[19] Boyd, P., and Schweber, L. (2012). "Variations in the mainstreaming of sustainability: a case study approach," in *Proceedings of the 28th ARCOM Conference,* 3–5 September 2012, Edinburgh, 1343–1354.

[20] Ye, K., and Xiong, B. (2011). "Corporate social performance of construction contractors in China: evidences from major firms," in *Proceedings of the 16th International Symposium on Advancement of Construction Management and Real Estate,* Chongqing, China

[21] Zhao, Z. Y., Zhao, X. J., Davidson, K., and Zuo, J. (2012). A corporate social responsibility indicator system for construction enterprises. *J. Clean. Prod.* 29–30, 277–289.

[22] Lingard, H. C., Francis, V., and Turner, M. (2010). The rhythms of project life: a longitudinal analysis of work hours and work–life experiences in construction. *Constr. Manage. Econ.* 28, 1085–1098.

[23] Loosemore, M., and Phua, F. (2011) *Responsible Corporate Strategy in Construction and Engineering: Doing the Right Thing?* Abingdon: Taylor & Francis.

[24] Post, J. E., Lawrence, A. T., and Weber, J. (1999). *Business and Society: Corporate Strategy, Public Policy, Ethics.* Boston, MA: Irwin/McGraw-Hill.

[25] Enimil, H. A., Agyemang, M. P., and Amankwa, A. O. (2012). Corporate social responsibility, a public relations gimmick? A case of Bharti Airtel Ghana Limited. Unpublished BA dissertation, Christian Service University College, Kumasi.

[26] Dahlsrud, A. (2008). How corporate social responsibility is defined: an analysis of 37 definitions. *Corp. Soc. Responsib. Environ. Manage.* 15, 1–13.

[27] Amponsah-Tawiah, K., and Dartey-Baah, K. (2016). "Corporate social responsibility in Ghana: a sectoral analysis," in *Corporate Social Responsibility in Sub-Saharan Africa,* eds S. Vertigaans et al. (Cham: Springer International Publishing), 189–216.

[28] Ocran, E. (2011). The effect of corporate social responsibility (CSR) on profitablility of multinational companies. A case study of Nestle Ghana Limited. Unpublished EMBA thesis, Kwame Nkrumah University of Science and Technology, Kumasi.

[29] Anku-Tsede, O., and Deffor, E. W. (2014). Corporate responsibility in Ghana: an overview of aspects of the regulatory regime. *Bus. Manage. Res.* 3, 31–41.

[30] Carroll, A. B. (1999). Corporate social responsibility: evolution of a definitional construct. *Bus. Soc.* 38, 268–295.

[31] Carroll, A. B. (2008). "A history of corporate social responsibility: concepts and practices," in *The Oxford Handbook of Corporate Social Responsibility*, eds A. Crane et al. (Oxford: Oxford University Press), 19–46.

[32] Carroll, A. B., and Shabana, K. M. (2010). The business case for corporate social responsibility: a review of concepts, research and practice. *Int. J. Manage. Rev.* 12, 85–104.

[33] Moura-Leite, R. C., and Padgett, R. C. (2011). Historical background of corporate social responsibility. *Soc. Responsib. J.* 7, 528–539.

[34] Lee, M. P. (2008). A review of the theories of corporate social responsibility: its evolutionary path and the road ahead. *Int. J. Manage. Rev.* 10, 53–73.

[35] Amponsah-Tawiah, K., and Dartey-Baah, K. (2011). Corporate social responsibility in Ghana. *Int. J. Bus. Soc. Sci.* 2, 107–112.

[36] Atuguba, R., and Dowuona-Hammond, C. (2006). *Corporate social responsibility in Ghana (Report submitted to Friedrich Ebert Foundation (FES)- Ghana)*. Available at: http://citeseerx.ist.psu.edu/viewdoc/download?doi=10.1.1.465.5190&rep=rep1&type=pdf

[37] Rockson, K. (2009). *Mainstreaming Corporate Social Responsibility (CSR) in Public Policy in Ghana: The Threats and Opportunities*. Available at: http://kwekurockson.com/docs/016.pdf

[38] Ofori, D. (2006). Business' corporate social responsibility: theory, opinion and evidence from Ghana. *Afr. J. Bus. Econ. Res.* 1, 11–40.

[39] Nyuur, R. B., Ofori, D. F., and Debrah, Y. (2014). Corporate social responsibility in sub-Saharan Africa: hindering and supporting factors. *Afr. J. Econ. Manage. Stud.* 5, 93–113.

[40] Pitt, M., Tucker, M., Riley, M., and Longden, J. (2009). Towards sustainable construction: promotion and best practices. *Constr. Innov. Inform. Process Manage.* 9, 201–224.

[41] Maignan, I., Ferrell, O. C., and Hult, G. T. M. (1999). Corporate citizenship: cultural antecedents and business benefits. *J. Acad. Market. Sci.* 27, 455–469.

[42] Carroll, A. B. (1979). A three-dimensional conceptual model of corporate social performance. *Acad. Manage. Rev.* 4, 497–505.

[43] Carroll, A. B. (1991). The pyramid of corporate social responsibility: toward the moral management of organizational stakeholders. *Bus. Horiz.* 34, 39–48.

[44] Moodley, K., Smith, N., and Christopher, N. P. (2008). Stakeholder matrix for ethical relationships in the construction industry. *Constr. Manage. Econ.* 26, 625–632.

[45] Yeboah, S. (2010). *Lifting the veil of "Corporate Social Responsibility" in Ghana.* Available at: https://www.ghanaweb.com/GhanaHomePage/NewsArchive/Lifting-the-Veil-of-Corporate-Social-Responsibility-in-Ghana-182648 [accessed May 24, 2010].

[46] Kuada, J., and Hinson, R. E. (2012). *Corporate Social Responsibility (CSR) Practices of Foreign and Local Companies in Ghana.* Malden, MA: Wiley Periodicals, Inc. doi: 10.1002/tie.21481

[47] Warnock, A. C. (2007). An overview of integrating instruments to achieve sustainable construction and buildings. *Manage. Environ. Qual.* 18, 427–441.

[48] Dartey-Baah, K., Amponsah-Tawiah, K., and Agbeibor, V. (2015). Corporate social responsibility in Ghana's National development. *Africa Today* 62, 71–92.

[49] Amponsah-Tawiah, K., Dartey-Baah, K., and Osam, K. (2015). Turning potential collision into cooperation in Ghana's oil industry. *Soc. Bus. Rev.* 10, 118–131.

[50] Jenkins, H. M. (2004). Corporate social responsibility and the mining industry: conflicts and constructs. *Corp. Soc. Responsib. Environ. Manage.* 11, 23–34.

[51] Ghana News Agency (2013). *Ghana Develops CSR Policy.* Available at: http://www.ccrgh.com/news/19-ghana-develops-csr-policy

[52] Ofori, D. (2007). Corporate social responsibility, myth, reality or empty rhetoric: Perspectives from the Ghana Stock Exchange. *Afr. Finance J.* 9, 53–68.

6

Work–Family Conciliation Policies: Answering to Corporate Social Responsibility – A Case Study

Adriana Faria and Carolina Feliciana Machado

Department of Management, School of Economics and Management, University of Minho, Braga, Portugal

Abstract

Living to and from the environment in which they are integrated, organizations are currently facing a huge challenge, such as to act in a socially responsible way. At the same time, the conciliation between work and family is a challenge that all workers face, as professional demands and family responsibilities are largely difficult to reconcile. In this sense, it is then important to implement good practices of conciliation within organizations so that the changes in both domains do not affect neither the work nor the family.

This chapter, which is part of an exploratory case study and based on a CSR context, addresses the conciliatory work–family policies implemented in a Portuguese company. Using the semi-structured interview as a way of obtaining data, in order to understand the adoption of conciliatory organizational practices between work and family life in a Portuguese medium company in the chemical sector, this chapter also aims to explore the resultant organizational and individual benefits.

The results demonstrate that the main factor for the implementation of such practices is the profit achieved by reducing labor conflicts, reducing costs, and improving work quality and organizational climate. It follows that the organizational benefits are productivity and the increase of the turnover.

At the individual level, there is a greater concentration and commitment in the work by the employee and a greater well-being having positive effects in the satisfaction of the employee. In this sense, and always in a perspective of social responsibility, the conciliatory work–family policies have an instrumental effect on employees' lives, since internal conciliation practices and mechanisms facilitate and support the responsibilities of family life with the objective of reducing the effects of these responsibilities on professional life.

6.1 Introduction

Diversity refers to the ways in which individuals differ, both in terms of personal bases and in terms of characteristics related to the organization. These characteristics are generally categorized as being visible (gender, age, race, appearance, and physical condition) and less visible (ethnicity, sexual and religious orientation, nationality, marital or parental status, political ideology, academic abilities, and social class) [1]. In this sense, diversity management requires the adoption of administrative measures that ensure that personal or team attributes are considered as resources to improve the organization's performance [2]. However, over the years, there are many barriers to its implementation, because implementing diversity management practices implies cultural and cognitive changes as well as a restructuring of human resources policies, namely, in the processes of career plans, in the valuation of the family, etc. (Gilbert et al., 1999, referred to by [2]).

In this chapter, it was intended, in a context of social responsibility, to study the management of diversity, focused on the theme of work–family conciliation within a Portuguese medium-sized enterprise. As the studies show that in Portugal, the issue of diversity is seen by companies merely as a logic of anti-discrimination rather than an inclusive strategy perspective [1], the purpose of this chapter was to carry out a case study in a company in the field of chemistry, where some work–family conciliation practices are implemented in order to understand:

- the main reasons that lead to the implementation of such practices,
- to explore the type of organizational and work–family facilitation policies that exist in the organization, and
- to identify the key benefits to the organization and individuals.

Having said this, the starting point that served as the guiding thread for this chapter was "What are the determining factors for the adoption of work–family conciliation practices in a medium-sized enterprise?"

The chapter is structured in four sections, the first one depicts the theoretical framework where the main analytical axes were approached around the theme, the second section describes the methodological options of the study, the third describes the case study and presents the analysis of the results and, finally, the conclusions of the study are presented.

6.2 Conciliatory Work–Family Organizational Policies

The concept of family refers to the sharing of a house by individuals who are linked to each other by biological ties, customs, or values. By its side, work is defined as all human activity whose objective is the production of goods or services for the maintenance of human life, implying joining to an organization that employs and compensates the worker for his or her contribution (Bourdieu, 1996 and Rothausen, 1999, referred to by [3]). In this sense, the concept of conciliation is defined as the equilibrium relationship established between work and family [4]).

The first interventions in the conciliation between work and the family were based on the need to support the mothers who work, not considering measures such as support for the elderly and children. But today, in a globalized world, family life cycles and career development stages are different from what happened years ago because women have been increasingly occupying managerial jobs, not neglecting family life and, in turn, men have come to assume more responsibilities and interest in parenting, as well as domestic and family work. Not forgetting, there is also the fact that individuals often have to provide support to dependents or the elderly.

In this sense, it has become necessary to reformulate organizational policies and procedures to support a career model that encompasses both spheres and not just work. However, there are difficulties in reconciling family and working life, which have been worsening, affecting equal opportunities and rights between men and women, both at the level of family life and in work situations. These difficulties in implementing the conciliation programs focus on cultural barriers and types of work organization, since the way of planning and organizing in enterprises does not take workers' needs into account, but rather a profit orientation [5].

According to the literature, there is no single theoretical model that clarifies the interaction between work and family. The classical perspective, defined as the theory of segmentation or separation, suggests that the domain of work and the family domain function independently and autonomously, without any mutual influences. Thus, individuals can achieve satisfaction and success with a domain without any influence from the other, as "the family is perceived as the domain of affectivity and expressiveness, while work is already seen as a domain of impersonality, competition, and instrumentality" ([3], p. 47). Thus, there is a conciliation between the two spheres because they function independently; however, this theory is criticized because the work–family relationship is seen to serve the interests of organizations.

The theory of conflict argues that individuals have limited energy and time, so that success or satisfaction in one domain (at work) implies sacrifices in the other domain (in the family). Therefore, there is the view that work and family are incompatible due to their different demands, responsibilities, expectations, and norms generating conflicts between the two spheres [3].

Instrumental theory, on the other hand, indicates that a given sphere (work) is the means by which one obtains what one wishes in the other sphere (the family). Thus, there is a conciliation between the two strands because the work emerges as a basis for success in the family sphere. Compensation theory also mentions that the inadequacies (dissatisfactions) felt in one domain are compensated by a greater investment in the other sphere of life, which leads to conciliation, that is, the rewards in a certain domain are insufficient and the individual seeks to obtain rewards of the same type in the other sphere [3].

Another theory is that of Spillover which is characterized by the existence of a reciprocal relationship between work and family in which one area of life influences the other in a positive or negative way. Emotions, predispositions, and behaviors are then transferred from one sphere to another, so that the existence of a boundary between the domains is not perceptible; however, although the causal relation can be established in both domains, processes cannot occur simultaneously [3].

From a more positive perspective comes the integrative theory that mentions the adjustment and the balance between professional and personal life that makes the domains fused into each other. This balance is not related to the equitable distribution of resources between roles, but to the individual's satisfaction with them [3].

Other investigations about the relation between professional and family life argue that also the labor or extra-labor roles affect this relation. Given these studies, it is concluded that either work or the family competes for an individual's limited time and energy, and that the performance of multiple roles ultimately results in stress and conflict symptoms (Goode, 1960, referred by [3]). In this context, boundary theory describes changes or transitions between roles as activities that involve crossing a set of boundaries, on a daily basis, where "boundaries" are lines of demarcation between different roles (or entities) and can take three distinct forms: physical, temporal, and psychological [3]. More concretely, individuals create and maintain boundaries as a way of ordering what surrounds them and in order to have a balance to reconcile the two spheres. Therefore, there are two central processes that affect the transitions between roles/borders: flexibility and permeability. Flexibility refers to the degree to which temporal and spatial boundaries are malleable, that is, they create flexible boundaries between the two spheres and do not interfere in performance between one and the other; permeability refers to the degree to which the performance of a given role allows it to be physically located in a given domain while being behaviorally and psychologically involved in another role.

From the development of theories on the subject, it is perceived that the need to reconcile work and family life is a feature of the contemporary labor market, being at the same time a policy closely linked to the organizational culture. Thus, the demand for harmonization of the labor and extra-labor spheres of workers "lives is a necessary condition for meeting the challenges posed by entrepreneurial competitiveness, as well as a means of guaranteeing the satisfaction of workers" needs [5]. Thus, the most recent reformulations of career–family reconciliation policies already include ascendants, elderly, and family members with disabilities or chronic diseases. There were also changes to European and international legislation on equal treatment of women and men in access to employment, vocational training and promotion, and working conditions, as well as on the balanced participation of women and men in work and family life and on childcare [6].

Nowadays, there are already companies that are flexible in the way of organizing work and managing time because they recognize that their employees have a family beyond their professional life, seeking, in this way, to implement organizational measures that allow these individuals to exercise with responsibility both papers. This type of business is defined as family-owned enterprises, which view the family as a "stakeholder" and implement a set of measures that favor harmonization between the two areas, allowing the employee to become a person and the society to develop in a more human way [5].

These firms define five types of work–life and family–friendly work practices, such as:

1. flexible time policies for employees, such as flexitime to meet family demands,
2. flexible space policies for employees, such as telecommuting or part-time work,
3. social benefit policies, such as life and health insurance extended to spouses and children of employees,
4. business policies of professional support to the worker, such as legal, financial, and psychological counseling of professional career development, taking into account the family situation of employees, and
5. family service policies, such as reduction of extra workload, leave for childcare or dependents (Family-owned businesses, 2010, referred to by [5]).

Studies show that applying these principles will enable companies to hire and retain the best professionals, will lead to improved image, and will help people to make the best of themselves [5]. Thus, the main organizational advantages of conciliation between the two domains are: increased productivity and reduced absenteeism, as it allows the planning of times more adjusted to the needs of the workers; deduction of the costs of creating services in the fiscal costs; and the enhancement of the company's image in the surrounding community and internationally, which contributes to the promotion of its products and to the increase of the business volume. In turn, at the individual level, they have positive effects on the individual's job satisfaction, on the time spent at work, as well as in the employee engagement and performance [5].

In the following sections, we intend to analyze and discuss how these policies are implemented in the context of a Portuguese medium-sized chemical company.

6.3 Methodological Options

This study had an exploratory character, of a qualitative nature, where it was intended to explore the type of conciliatory organizational work–family practices of a particular company. The study is based on the phenomenological paradigm because it was sought to understand the theme from the concrete reality, and the research method used was the case study [7] that allowed to explore and capture specific characteristics of how and why are implemented the conciliatory work–family practices in a real context of a company. Thus, the object of analysis that composed the study was a medium-sized family company, of the chemical industry sector. The type of sampling used was intentional non-random sampling because the selection of the analysis unit in question is characterized by close contact with the researcher.

The starting point of the study was "*What are the determining factors for the adoption of work–family-friendly organizational practices in a medium-sized enterprise?*" In what concerns the main objectives, this study looks to explore the type of organizational work–family reconciliation practices implemented by the company; to explore the reasons for the implementation of these practices; and, finally, to identify the main benefits of this conciliation for employees and organization. The variables of the study were the work–family reconciliation practices implemented in the organization (dependent variable) and the determinants of the implementation of these organizational policies (independent variable).

In order to acquire and deepen the notions and knowledge of this subject, a theoretical review was carried out and a semi-structured interview was elaborated to carry out a more in-depth study of the elements of analysis. The interview consisted of guiding questions that served as a guide to the researcher and was administered to a manager of the company, on April 8, 2017, safeguarding a flexible and fluid development. The technique of information processing was the analysis of thematic content elapsed from the themes axis of the interview script.

6.4 Case Study: Analysis and Discussion of Results

This section aims to present the data of the case study and the analysis and discussion of the results based on the interview script that resulted from a certain set of themes identified as relevant to the study. For reasons of confidentiality, the name of the company and the interviewee will not be disclosed.

6.4.1 Company Characterization

The company under study is a medium-sized family-owned company in the chemical industry that produces and markets resin derivatives to several countries. Its leading suppliers are the European countries in the industry.

The company transforms resin for use in tires, printing inks, varnishes, adhesives, etc. Its mission is to offer customers the best resin in the market, guaranteeing the highest level of quality and customer satisfaction. The vision is to achieve operational excellence in all activities and the organizational policy is characterized by the requirement for quality, high levels of employee safety and products, and responsible environmental protection (development of ecological resins).

The organization was considered an SME leading excellence in the years 2010, 2011, and 2012 and holds ISO9001 quality certification, OSHAS18001 safety system certification inspected by external bodies, and ISO14001 environment certification in which it promotes awareness, practice, and training for employees, of consistent improvements in practices with environmental impact and prevention of environmental contamination.

The company is characterized by:

- an administrative department, where the administrators, accounting, financial sector, commercials, vendors, and other senior staff are located,
- a human resources management department with three employees dedicated to business strategy and administrative and operational human resource management,
- a internal maintenance department, production, a quality control office, a chemical production laboratory made up of chemical analysts, and
- a research and development department made up of chemical engineers dedicated to the discovery of ecological and sustainable solutions to protect the environment, provide technical support to the market and optimize laboratory products for the industrial scale.

6.4.2 Human Resource Characterization

The company has 140 employees, of whom 40 are women in the research, quality control, and administration sectors. In these sectors, the vast majority of top managers and middle managers are women. In the production laboratory, there are women with the function of technical analysts and chemical researchers and there is an intermediate position (Chemical production engineer). There is one responsible for the quality system, one industrial director and one technician responsible for each sector. Human resources are characterized by Portuguese, Ukrainian, Russian, and other Eastern European nationalities (currently 20 employees, men, and women, of foreign nationality) and has a team of people with disabilities to deal with gardening.

6.4.3 Human Resource Management Practices

Recruitment and selection processes are carried out by the company, formalized through advertisements in newspapers, employment center, partnerships with institutions, and university. The most used techniques are the analysis of curricula and interviews coordinated by specialists in the field. Required qualifications are technical professionals and university degrees in the areas of chemistry (industrial and laboratory), forestry, architecture, and management. The company enters into indefinite contracts with all employees, including trainees who, after 9 or 12 months, enter, effectively, to the company.

The process of reception and integration results from a document where the company sectors, the activities, and the hierarchical heads are designated, having a visit to the different sectors of the organization. The training process involves all employees, takes place throughout the year, and focuses on hygiene, health, and safety at work, first aid, fire, innovation, professional retraining, and improvement/updating of skills. There is an initial training for the new employees at the job site, for a month, and with direct supervisor supervision. All training takes place in the company, in the theoretical–practical modality, during working hours, with specialized external trainers and with evaluation of initial and final training. The survey of training needs arises from the needs of each task analyzed in the annual planning, by the employees' performance appraisal, and by the quality requirements or technological requirements.

With regard to compensation' management, it is defined by the function or job position/task. It should be noted that compensations of all functions are slightly above the national average, which makes this management practice a retention factor for employees in the company.

Performance appraisal is individual, occurs annually, on a regular basis, following a set of defined criteria/requirements through compliance with hygiene, health, and safety standards at work; labor process and quality standards; and an evaluation is made by the intermediate/direct manager to the employee. There is a fairly significant annual performance award for all employees, which results from compliance with hygiene, safety, and health at work rules, which is another employee retention factor, as well as recognition of professional performance through the communication between managers and employees and administrators.

6.4.4 Diversity Management

Individuals of foreign nationality have been in the company for 5 years; however, 6 years ago, six individuals with the same foreign nationalities entered. According to the interviewee, *"the first Ukrainians, Russians [...] sought employment on their own initiative and after a period of experience were hired for indefinite term in the company just like us."* It is then realized that these hiring were not the result of the human resources planning process, but resulted from one of the values of the company, which is not to exclude and discriminate.

The organization has continually adapted, that is, individuals have been welcomed into the company gradually and integrated into the work teams progressively. There was a professional retraining of the training of these individuals and a flexibility of communication, such as the translation of workbooks and norms in foreign language and the translation by work colleagues in the daily tasks. The company also provided external training to help in learning the Portuguese language, because *"language was the greatest difficulty felt by all."* (Interviewee)

These individual' families came to our country after their establishment in Portugal. To facilitate this process and ensure a greater conciliation of work and family life, the company has created residential housing for these foreign individuals and their families. After this process, and to date, the organization has hired more than 30–40 individuals (men and women) with these nationalities, increasing the residential housing for these employees and their families, adapting the cafeteria with the installation of microwaves to the meals that foreign individuals bring from home. At the beginning, recruitment was done through *"give the word to foreign relatives/friends"* (Interviewee); however, the organization changed the recruitment policy in order to incorporate diversity management and, currently, human resources' planning incorporates diversity and has a link to strategic business planning. The training process was also reformulated to facilitate the integration of these individuals into work activities.

"As there is an intention to recruit individuals with these nationalities, it was necessary to formalize procedures and processes and change internal records and even company conditions. The training for more practical modality and with translations for a faster insertion of them in the work and that benefited, also, the internal Portuguese employees, was reformulated. Today, foreign individuals who enter already know Portuguese, even if the minimum, but 15 years ago no. The internal regulations were changed and

today there is already a person in the management of human resources linked to the company's strategy and business and that incorporates this type of management." (Interviewee)

According to the results of the interview, the main benefits of this diversity incorporation in the company are:

- the more diversified productive workforce, flexibility, creativity, an organizational culture open to change and difference,
- and the sharing of work methods, different experience, and ideas that lead to innovation and individual growth and values the company's image in the surrounding community and internationally.

In another context, the existence of individuals with disabilities to provide gardening services results from a partnership between the organization and a disability institution that welcomes individuals to a center of occupational activities, having positive effects for both parties. However, more than the productive component, the daily personal and professional development of individuals with disabilities matters, because, and as the interviewee indicates, *"they are good workers and we can see progress over time, the autonomy they earn and the motivation they have because they see that they are contributing to something, they see the results, and feel useful."* (Interviewee)

According to the studies, social responsibility practices most developed by national companies are based on the economic dimension and the internal social dimension, being also high the practices in the environmental dimension and less mentioned in the external social dimension. In the case in question, it can be seen that the company's social responsibility, in addition to the sustainability and economic component, is committed to the external social dimension by building a green park of 13 hectares of area for social, recreational, and employee and family use. However, the studies claim that there is a valuation of practices related to the fulfillment of legal obligations and there is a gap in the management of organizational change and the participation of workers in the process (Santos, 2006, referred by [1]). In our case study, it was verified that there is a real gap in the management of the organizational change. Indeed, although the intention of the management was to incorporate the diversity in the company and a participation and collaboration of all in its implementation, there wasn't, according to the interviewee, a formalization of change and of diversity management communication nor was a specific training on diversity management for employees developed.

One can see that the company develops a diversity management according to the Melting Pot Model where equality arises based on difference and where everybody equal treatment is ensured. This is observed from the inclusion of individuals of other nationalities and from the initial changes that occurred in the company with the purpose of adapting the services and conditions to accommodate the employees of different nationalities. Subsequently, the organization has already changed its way of managing diversity, attempting to adopt a model of diversity management according to the Cultural Mosaic Model, where equality is based on difference and where top management intends to recruit and retain a mix heterogeneity of employees who have competencies that generate value to the company and which results in a competitive advantage. There is, therefore, an inclusive diversity management, even if it is at an initial level and where the organization has already experienced the benefits derived therefrom, as previously mentioned.

6.4.5 Organizational Policies for Work–Family Conciliation

With regard to the dimension *(i) work flexibility*, this flexibility must satisfy both employers and workers.

The administrative department works 40 h a week and the entire production, and the maintenance and research department work 24 h a day, distributed in three 8-h shifts, with shift heads and work teams. There are no overtime hours in all sectors, the company closes in December and August for vacations, and the work is carried out in teams which allows to make necessary exchanges.

Regarding the policy of work flexibility, it is observed that the organization works by rotating shifts that allow a labor flexibility. Nevertheless, it is perceived that this flexibility had for greater purpose the management of productivity and obtaining greater profits.

"The company worked in normal daytime and weekly hours, but with the growth of the business, and the increased of production, it was necessary to restructure production in order to be able to respond to the surplus volume of orders, production, etc. As we have European suppliers in the same field and with the full operation of the company in rotating shifts, it was decided to experience this situation here to respond to the productivity we were feeling and feel until today. It was complicated, it was necessary to restructure teams, plan the best times and conditions, instill this culture in the company and believe that employees would join. Some left voluntarily and those who

stayed here were rewarded and benefited from better conditions, increased salaries, bonuses, etc. Later, we managed to create the best shifts according to our needs and theirs, there was a transfer from both parties and it works until today; we only stop in August and at Christmas for maintenance." (Interviewee)

Despite this factor, we see advantages in this flexible schedule for employees.

"Shift work allows you to always have a morning, afternoon or evening available to reconcile rest and other personal and family tasks from day to day. There are always 16 free hours that are taken advantage of differently, from professional daytime work, which is divided into 4 h in the morning and afternoon, having, always, weekly breaks, where the smallest has a duration of 32 h. Having the annual calendar of the already defined schedule at the beginning of the year, allow to plan their life more in advance and concretely." (Interviewee)

Comparatively, the dimension *(ii) space flexibility policies for workers* is not applied because there is no existence of telework or partial work.

Regarding the dimension *(iii) policy of social services or supports*, the organization has a canteen/cafeteria with reduced price meals for all staff and microwave to warm meals/kitchens of the employees. The food allowance is above €5 and is assigned by food card to all employees. The company's Christmas party is similarly held in the cafeteria where a gift is distributed to the children of the employees. Healthcare support is also available, there is a doctor and nurse present in the organization, and a health insurance, extended to the spouse and children, is offered to all employees. There is also a discount for all employees with a laboratory for laboratory analysis, ophthalmology, and otolaryngology.

In addition to social support, there are, jointly, services or equipment to support children within the facilities. The day care/kindergarten inside the facilities has three to five educational aids, with the extended operation, free for all the children of the employees, being also open to children of public outside with reduced prices. In total, there are already 30 children and has interactive and resting rooms, bathrooms, and outdoor play space with teaching equipment.

In the policy of social services and supports, the studies presented in Section 6.1 of this chapter show that day care support in the workplace can be a fundamental means of guaranteeing the possibilities of reconciling work and family life and, in this way, preventing one of the factors potentially more discriminator between men and women [6]. The main advantages for working parents are:

"It is easy to leave the children in the place until later, flexibility of transport and time because the place of work and day care is the same not being so much inconvenience. It is also less expensive in the financial level even because the service is free, contributing for not have the stress, the rush and worry where to leave the children. The compatibility of the schedules is also easier to manage." (Interviewee)

The interviewee's speech confirms Torres and Silva's study (1998, referred to by [4]) that shows that there is a difficulty in reconciling the issue of childcare on the return to work, with the options divided between the help of family members or the kindergartens that parents have to pay. There are also empirical studies that identify three characteristics derived from the conflict between work and family that can sometimes be incompatible with the role to play in the family. They are: the scarcity of time, in which time dedicated to work can absorb the individual in such a way that makes it impossible to give attention to the family; the dedication and effort placed on the performance of one of the roles that creates stress will affect the individual's predisposition to performance in another role and, finally, the demands of behavior that a certain role may require in the sphere of work (Greenhaus and Beutell, 1985, referred to as [4]). So, the research highlights that, in order to reduce stress at work, organizations must implement policies that allow the creation of family support services and, on the other hand, allow employees greater control over working time and greater flexibility in terms of schedules (Appelbaum et al., 2005, referred to by [4]).

In the case under study, the company implements social/family support policies to reduce the effect that family responsibilities have on working life. One of the main objectives of this implementation is for employees to be more relaxed about child care, having a better coordination of their work schedules with family tasks reducing the stress and pressure that this represents in everyday life.

On the other hand, support for dependents or elderly people is already a more difficult factor to manage because companies are easier to have support for children than for the elderly.

"Having support for the children in the internal structures of the company is easier than giving support to the elderly or dependent, but of course there is flexibility of work and schedule, we have this in consideration and this attention as much as possible. And there are financial support so that in such

cases, the employees can seek help in other institutions and even counseling or the company is the intermediary in the process." (Interviewee)

There is also the promotion of sporting activities, through the existence of a football field with athletics track available to all employees and where activities such as games and daily races are performed. Sports services have as their main objective *"to promote healthy relationships, improve communication between colleagues and managers, and create an easy access space that allows employees an abstraction and relaxation."* (Interviewee)

The dimension *(iv) family services policies* is characterized by the existence of maternity leave, parental, breast-feeding, and leave to care for children and dependents.

In the organization under study, in case of pregnancy, all women (with the exception of the administrative sector) are exempted from work, in order to protect the embryo and the woman, from the moment they announce their pregnancy to the company, because as it is of a chemical industry sector workers are subject to risk factors.

For all sectors, maternity leave lasts for 120 days or 150 days in a row, according to the law, or the license is shared with the husband, which can be increased by another 30 days. The breast-feeding license is of two distinct periods of up to 1 h each, without loss of remuneration or any benefits, for all female employees, from all sectors, with the possibility of choosing the exempted hours. There is also the exemption for prenatal consultations for mothers and fathers without a discount in compensation.

As far as maternity leave is concerned, to date all female employees have enjoyed full maternity leave and free days, as well as breast-feeding, with support and mutual assistance from colleagues. In the case of men, there is also the promotion of parental leave.

"In the case of women all enjoy the total licenses and free days and in the case of men, it is tried to instill the culture of parenthood and the boss is the first to wonder why men do not want the dispensation if the company gives the parental license or why the man does not make the shared license. In my case, as I belong to the management team, it is not so easy or practical to replace. But with the proper planning it is done and with mutual help of everyone it is made, shifts were shifted, etc. and I enjoyed the parental leave, they were 10 days more a week and I enjoyed those free days for consultations and child care. In the past, men did not have the right to keep up with the growth of their son so close. But today, it is possible under the law, and therefore the company promotes this reconciliation for internal and external well-being

and among all of us there is help because we all have families and we care about work and family and then when there is no conciliation it is difficult to work properly. There is a spirit of camaraderie. In this issue, there were never problems, we always get substitution of each other and we help each other to fill some rooks and it is very good." (Interviewee)

The parent's exclusive parental leave is of 25 business days, followed or interpolated, concurrently with the mother's license, paid in full and there is the possibility of 120 days of initial parental leave paid in full. There are also free days that allow workers greater flexibility in family support, such as free days to children care. The company has flexibility for family assistance or personal matters to all employees, by informal notice, without loss of salary or need for compensation of hours. All benefits are defined in the internal policy regulation. It is relevant to emphasize that in all cases, jobs, as well as working conditions, are ensured for all sectors.

According to Bailyn (1997, referred to by [4]) managers should see the conciliation of work and family not individually, but in a systemic and integrated way as part of the organization and the work culture, in order to create a work environment that frees up the potential of all employees. In this context, it is important to positively highlight the acceptance and support of colleagues and the organizational culture of support to parenthood and maternity, as well as the promotion of equal opportunities for all in the exercise of parenting, encouraging the sharing of parental leave, without prejudice the mother' exclusive rights.

Regarding the dimension *(v) corporate policies of professional support to the worker*, the organization makes possible financial support through individual loan without interest, to all the employees.

Finally, with regard to the reasons for the implementation of the organizational policies for reconciling employment and family, the main factor was the reduction of costs achieved through the reduction of labor conflicts, the employees' quality of life, individual well-being, and harmony that results from the conciliation policies.

"It cannot be denied that one of the objectives is profit through cost reduction because these measures facilitate the individual's well-being and concentration at work as being freed from some family concerns they are more focused, generating good climate, performance, and productivity. Less stress

with daily problems is generated. There is quality of life as we try to reduce the weight of external factors to the work that influence the day to day work, such as not have where to leave the children, raids, and stress, to pay for expensive childcare services and this is all worries in anyone's head. So, I think that are linked goals and benefits because the good internal climate is only possible with the employees well-being, that is linked to internal and external factors. There are factors that we do not control, but there are others that can reduce their effect on people's daily life and this is what we try to do, to reconcile the two sides to have harmony, and family and organizational tranquility, because if the workers have family concerns they are not able to work well daily." (Interviewee)

It is concluded, therefore, that the determining factors for the implementation of organizational policies of work–family conciliation are costs reduction, productivity increase, and, consequently, profit. Given the discourse, it is perceived that the reduction of costs comes from the reduction of the labor conflicts that arise from the workers well-being, the good organizational climate, and the work–family conciliation that reduces the employees' family concerns, which contributes to a better commitment and performance, and increased productivity. This corroborates the empirical studies that support the idea under which the workers who feel comfortable in their personal lives and confident in their place of work tend to perform better in employment [4].

6.5 Final Considerations

This chapter summarizes the main conclusions of the study, their contributions, and limitations.

The study aimed to analyze the policies of work–family conciliation from the real context of a company, focusing on the reasons and the causes of this implementation and the benefits arising therefrom.

In the light of all the foregoing, it is understood that the work and family conciliation policy in the organization under study has an instrumental effect on individuals in which the professional area is the basis for family success and conciliation. In other words, the organization has internal practices and mechanisms that facilitate and support the family life responsibilities of employees in order to reduce the effects of these responsibilities on professional life. Therefore, the professional domain emerges as a benefit to the family domain in which employees enjoy organizational measures that lead to a conciliation between work and family.

Emphasis is placed on the company's relevance given to human resources diversity, with the intention of continuing to recruit and retain diverse cultures. To this, it will be important to formally promote diversity and prepare individuals for continuous change so that management becomes more effective and assesses or controls its effects.

One limitation of the study was the shortage of time that did not allow formal interviewing of other types of employees of the company that do not belong to the higher functions and it was not possible to consult internal documentation related to the measures adopted. However, it was possible to approach some workers in informal conversation and confirm some speeches of the interviewee. It would be interesting in future investigations to analyze this theme more specifically focused on the benefits of this conciliation in the daily routine of workers.

References

[1] Gomes, S., Augusto, C., Lopes, M., and Ribeiro, V. (2008). *A Gestão de Diversidade em Pequenas e Médias Empresas Europeias.* Lisboa: Parceria de desenvolvimento – Respons & Ability – Investindo na diversidade, Iniciativa Comunitària EQUAL.

[2] Alves, M., and Silva, L. (2004). A crítica da gestão da diversidade nas organizações. *Rev. Análise Empres.* 44, 20–29.

[3] Santos, G. G. (2011). *Desenvolvimento De Carreira: Uma Análise Centrada na Relação Entre o Trabalho e a Família.* Lisboa: RH Editora.

[4] Costa, J. (2012). *Práticas de Conciliação entre o Trabalho e a Família: Um Estudo Exploratório.* Dissertação de Mestrado em Gestão, Escola de economia e gestão, Universidade do Minho, Braga.

[5] Hamid, F. (2012). *Práticas De Conciliação Trabalho/Família em Organizações de Excelência.* Dissertação de mestrado em Gestão, Universidade de Coimbra, Coimbra.

[6] Guerreiro, M. and Pereira, I. (2006). Responsabilidade social das empresas, igualdade e conciliação, trabalho-família – Experiências do prémio igualdade é qualidade, CITE. *Estudos* 5, 1–110.

[7] Yin, K. (2001). *Estudo De Caso: Planeamento E Métodos*, 2nd Edn. Porto Alegre: Bookman Editora.

Index

About the Editors

Carolina Machado received her PhD degree in Management Sciences (Organizational and Politics Management area/Human Resources Management) from the University of Minho in 1999, and Master degree in Management (Strategic Human Resource Management) from Technical University of Lisbon in 1994. Teaching in the Human Resources Management subjects since 1989 at University of Minho, she is since 2004 Associated Professor, with experience and research interest areas in the field of Human Resource Management, International Human Resource Management, Human Resource Management in Small and Medium Enterprises, Training and Development, Management Change and Knowledge Management. She is Head of Human Resources Management Work Group at University of Minho, as well as Chief Editor of the International Journal of Applied Management Sciences and Engineering (IJAMSE), Guest Editor of journals, books Editor and book Series Editor, as well as reviewer in different international prestigious journals. In addition, she has also published both as editor/co-editor and as author/co-author several books, book chapters and articles in journals and conferences.

João Paulo Davim received the Ph.D. degree in Mechanical Engineering in 1997, the M.Sc. degree in Mechanical Engineering (materials and manufacturing processes) in 1991, the Mechanical Engineer degree (MEng-5 years) in 1986, from the University of Porto (FEUP), the Aggregate title (Full Habilitation) from the University of Coimbra in 2005 and the D.Sc. from London Metropolitan University in 2013. He is Eur Ing by FEANI-Brussels and Senior Chartered Engineer by the Portuguese Institution of Engineers with a MBA and Specialist title in Engineering and Industrial Management. Currently, he is Professor at the Department of Mechanical Engineering of the University of Aveiro, Portugal. He has more than 30 years of teaching and research experience in Manufacturing, Materials and Mechanical Engineering with special emphasis in Machining & Tribology. He has also interest in Management & Industrial Engineering and Higher Education for Sustainability & Engineering Education. He has guided large numbers of postdoc, Ph.D. and masters students. He has received several scientific awards. He has worked as evaluator of projects for international research agencies as well as examiner of Ph.D. thesis for many universities. He is the Editor in Chief of several international journals, Guest Editor of journals, books Editor, book Series Editor and Scientific Advisory for many international journals and conferences. Presently, he is an Editorial Board member of 25 international journals and acts as reviewer for more than 80 prestigious Web of Science journals. In addition, he has also published as editor (and co-editor) more than 100 books and as author (and co-author) more than 10 books, 70 book chapters and 400 articles in journals and conferences (more than 200 articles in journals indexed in Web of Science core collection/h-index 44+/5500+ citations and SCOPUS/h-index 52+/8000+ citations).